CONTENTS

PRACTICALITIES

- Understanding the calendar 54
- Calculating world times 80
- Index 112
- How to use the calendar (tear out this page) 113

UNDERSTANDING THE MOON

- The waxing and waning Moon 2
- The ascending and descending Moon 4
- The distance from the Earth to the Moon 6
- Lunar nodes/Planetary nodes/Eclipses 7
- Zodiac signs and constellations 8
- The Moon in the zodiac signs 9
- The Moon in the constellations 10
- When signs and constellations are aligned 10
- Planetary aspects 11
- The Red Moon/May Frost/Indian Summer 15
- Tides 16
- Chinese seasons 18
- Geobiology 19

WORK

- Gardeni
- Seeds
- Daily rhy
- Potatoes
- Horticultu 24
- Cultivating trees and shrubs 25
- Harvesting 27
- Plant-based decoctions 29
- Fungicides and insecticides 30
- Compost 31
- Crop rotation 32
- Agriculture 33
- Hay 34
- Cereal crops 35
- Animal husbandry 36
- Beekeeping 38
- Winegrowing 39
- Cider making/Beer making/Forestry 40
- Miscellaneous 42
- Plan of your garden 109
- Companion planting 110

LIVING WITH THE MOON

- Influence of the phases 44
- Harvesting plants/For making infusions/ Medicinal plants 45
- Hairdressing 46
- Depilation 47
- Skincare/Warts 48
- Nails/Corns and callouses 49
- Fasting/Detoxing/Treating worms 49
- Eating 50

LUNAR CALENDAR

- The calendar 56
- Gardening notes 81

UNDERSTANDING THE MOON

The waxing and waning Moon

This is one of the most familiar features of the Moon as viewed from the Earth. What we are seeing is the Moon's monthly orbit around the Earth (aka the synodic revolution), which starts with the New Moon and takes 29 days, 12 hours and 44 minutes. When new, the Moon is positioned exactly between the Earth and the Sun – thus, the illuminated area of its surface is not visible from Earth. This phase is represented on the diagram below by a black circle. As the Moon progresses on its orbit, it reflects a crescent of light that expands until it is seen from Earth as a luminous disc – the Full Moon. In this phase, the Moon is on the opposite side of the Earth to the Sun. From that point on, the illuminated area decreases until the Moon renews its orbit once more.

The waxing Moon

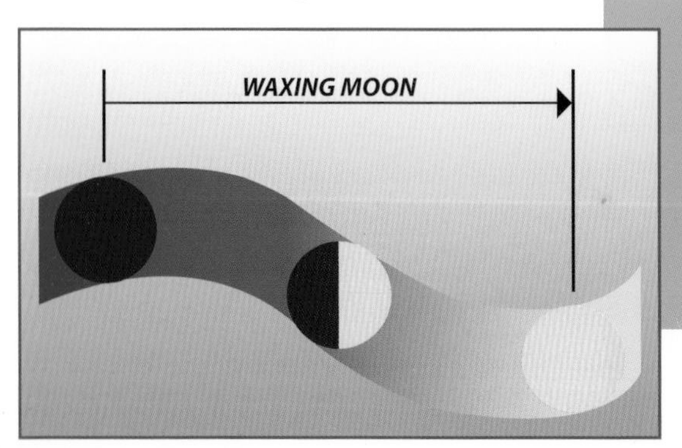

The Moon 'waxes' during the phase from New Moon to Full Moon – the illuminated area increases in size every day. On the calendar (pages 56-79), the colour of the blue band also becomes lighter as the Moon waxes.

In the night sky of the northern hemisphere, a quick glance will tell you if the Moon is waxing – the illuminated area is shaped like a crescent, which, if you were to add an imaginary line to the left of it, would resemble the letter 'p'. In the southern hemisphere, the illuminated area is also expanding but is seen the opposite way round, while in equatorial regions the crescent appears to be lying on its back.

This phase is indicated on the lunar calendar (pages 56-79) between the New Moon and the Full Moon.

Plants increase in vitality with moonlight and as the Full Moon approaches, their resistance to parasites and diseases increases. Fruits and vegetables harvested at this time store well and impart more vitality when eaten, while cut flowers last longer in a vase. Silage and mown hay are of better quality, compost is warmer and animals are less anxious when there are people around.

The waning Moon

The Moon 'wanes' during the phase between the Full Moon and the next New Moon. Every night the illuminated area becomes smaller. On the calendar (pages 56-79), the colour of the blue band darkens as the Moon wanes.

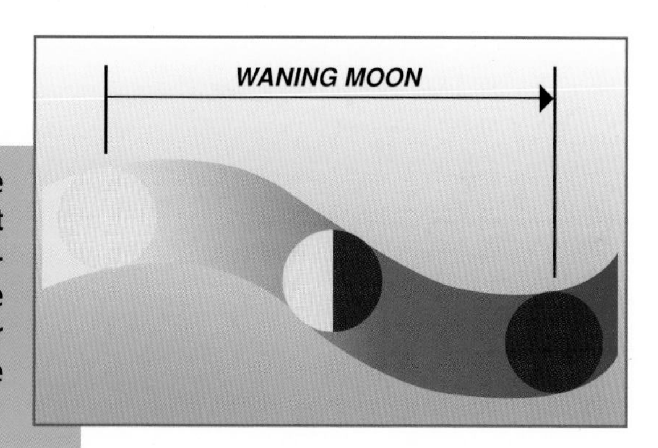

In the sky of the northern hemisphere, the illuminated area of the waning Moon reverts to a crescent shape, but now if you were to add an imaginary line to the right of it, it would resemble the letter 'd'.

This phase lasts from the Full Moon to the next New Moon.

As the moonlight decreases, so does the vigour of plants, although their specific energy is increased – colours, scents and tastes are more perceptible during this phase, and nutritional and medicinal properties are more pronounced. However, it is more difficult to store harvested crops in their natural state and this phase is more suitable for preserving foods, making jams and bottling wine. Insecticides and fongicides are more effective.

The ascending and descending Moon

Many people think that an ascending Moon is the same as a waxing one when in fact they are totally different. The Moon can, for example, wax and descend at the same time. The path of the ascending and descending Moon is similar to the progress of the Sun during the course of the year. In the northern hemisphere, the Sun rises in the south-east and sets in the south-west at the winter solstice, toward the end of December. The arc that it describes in the sky is very short and at noon it is very low on the southern horizon. The closer we are to the summer solstice at the end of June, the nearer to the north-east the Sun rises and the nearer to the north-west it sets. Its arc is much longer at this time and at noon the Sun is very close to its zenith. The sun is therefore ascending during the six months between the winter and summer solstices and descending during the six months between the summer and winter solstices. The Moon also ascends and descends, but over a period of 27 days, 7 hours and 43 minutes, known as the **periodic lunar cycle**. In the northern hemisphere, the Moon ascends in the sky and then descends again, while in the southern hemisphere the reverse takes place.

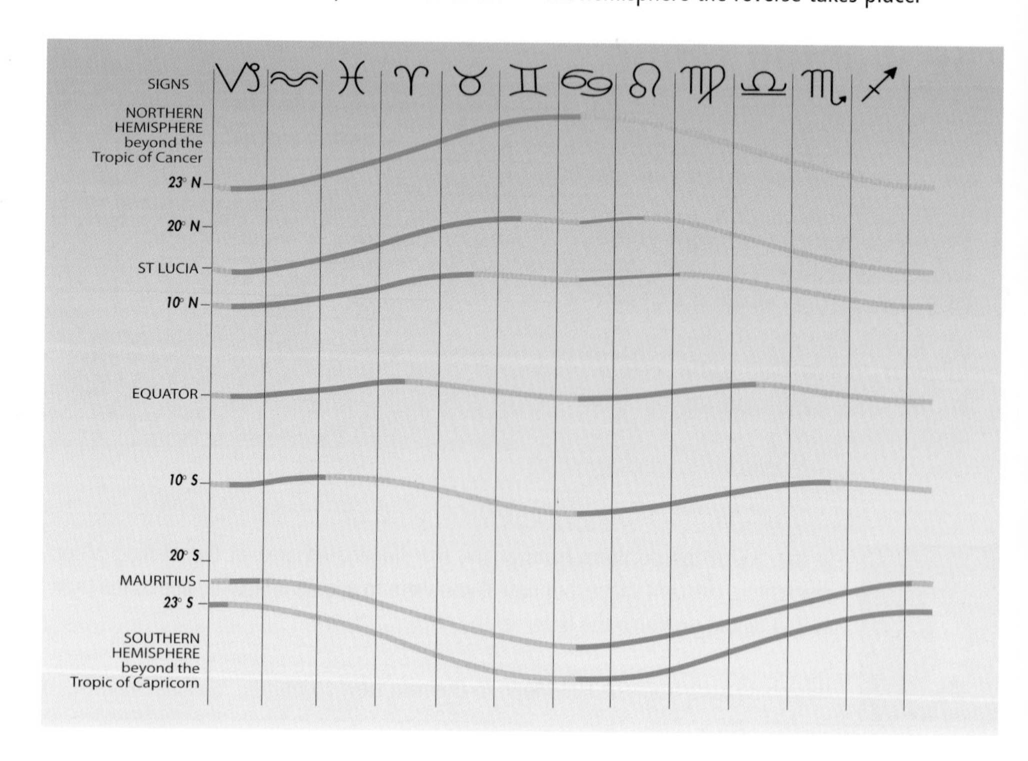

On the calendar on pages 56-79, the narrow coloured band beneath the date band shows the path of the ascending and descending Moon in the northern hemisphere as far as the Tropic of Cancer (to determine the path in the southern hemisphere, south of the Tropic of Capricorn, simply invert the colours). Between the two tropics, there is a regular pattern of inversion (see diagram above).

The influences of the ascending and descending Moon and Sun grow steadily weaker the nearer we get to the Equator. Beyond the tropics, the forces that result from such movements are linked with three factors: the ascending/descending Moon, the waxing/waning Moon and the tides (see page 16). Any of these factors can predominate – for example, in Europe, grafting is best done when the Moon is ascending. However, the nearer you are to the Equator, the more the influence of the ascending/ descending Moon diminishes, while that of the two other factors increases. For instance, nearer the Equator it is more beneficial to graft plants when the Moon is waxing than when it is ascending.

In the northern and southern hemispheres, in regions beyond the tropics, the ascending/descending Moon has a very strong influence in certain areas of the lunar calendar. In equatorial regions, on the other hand, it is the waxing/ waning Moon and the influences of the tide that have greater influence.

The Moon is ascending when, every night, its orbit is higher than the night before. Look at the Moon, ignoring the stars, for two consecutive nights to work out whether it is ascending or descending. On the first night, establish its position in the sky as accurately as possible. A little later on the following night, the Moon will pass through the same vertical once more. If it passes through at a greater height, you will know that it is ascending.

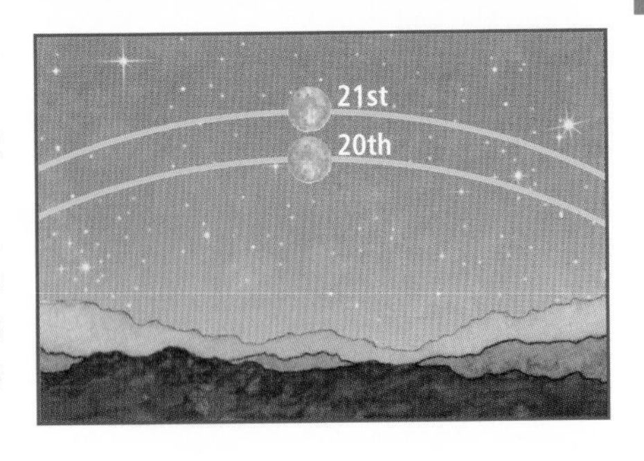

The ascending Moon

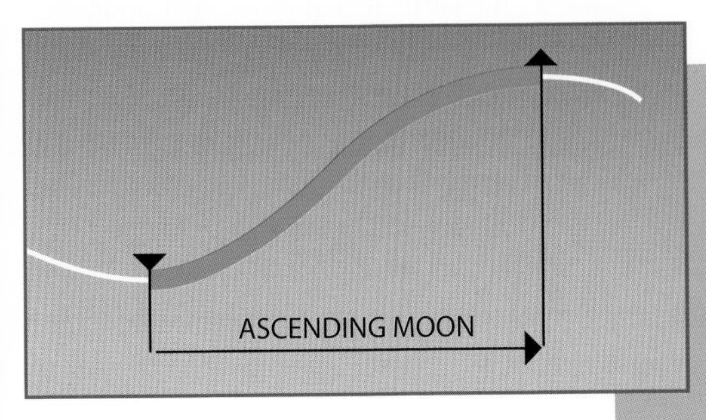

On the calendar (pages 56-79), the Moon is ascending when it travels from its lowest point (▼) to its highest (▲). This continual oscillation is shown in the illustration. When the curve rises, the Moon is ascending.

The fluids within plants rise and fall with the Moon. When the Moon is ascending, plants contain more sap so there is more activity in their aerial parts (those parts above ground). This is a good time, for example, to cut scions, **to graft plants and to harvest fruit** with a high juice content, as well as to collect sap from silver birches, etc. However, it is better to avoid pruning trees or cutting plants for drying at this time. Cutting lawns when the Moon is ascending tends to encourage plenty of growth; it is a good time to aerate them.

The descending Moon

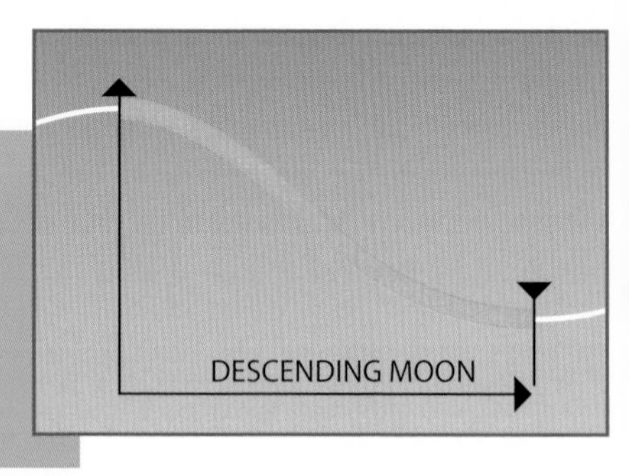

On the calendar (pages 56-79), the Moon is descending when it travels from its highest point (▲) to its lowest (▼). When the curve falls, the Moon is descending.

When the moon is descending, the flow of fluids in plants descends also and growth occurs mainly in the roots. This is a good time to **harvest root crops** or the aerial parts of a plant that you want to dry quickly, and also for **pruning, pricking out, re-potting, ploughing,** spreading **compost or manure** and **cutting wood**. The grass of lawns that are cut at this time forms stronger roots and anchors the soil better.

The distance from the Earth to the Moon

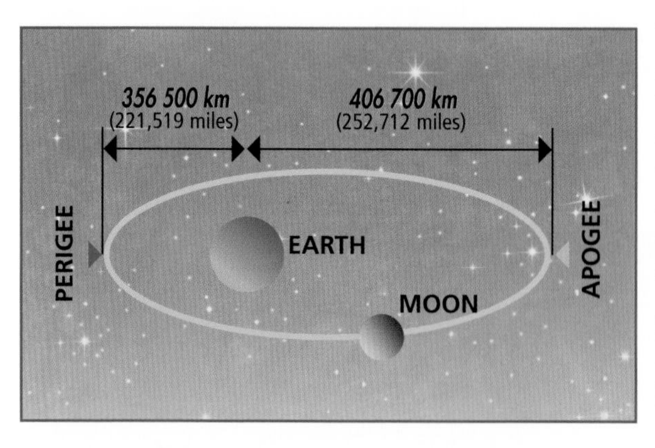

The Moon moves around an elliptical orbit, with the Earth as one of its foci. This means that the distance between the Earth and Moon is always changing. When the Moon is closest to the Earth, it is at its perigee, and when it is furthest away, it is at its apogee. The progress of the Moon from one perigee to the next is known as the **anomalistic lunar cycle** and takes 27 days, 13 hours and 18 minutes.

The Moon's effects on plants increases when it nears the Earth and its influence is, therefore, strongest around the time of the perigee. It is advisable to avoid all work involving soil and plants on the day of the perigee itself.

The Moon is closer to the Earth at the perigee; it looks larger at this time and, conversely, smaller at the apogee. In the calendar on pages 56-79, the Moon is shown at its largest near the perigee and at its smallest near the apogee, set in a band that is similarly widest at the perigee and narrowest at the apogee. The point of the perigee is indicated by the letter P and of the apogee by the letter A.

Lunar nodes

The Earth moves in an elliptical orbit, with the Sun as one of its foci. The plane of this ellipse is called the **ecliptic**. The Moon also moves around an elliptical orbit, with the Earth as a focus. The plane of this ellipse is at an angle of 5° 9′ to the ecliptic. As it moves around the Earth, the Moon therefore crosses the ecliptic twice: once when descending – this is the **descending node** (☋), and once when ascending – which is the **ascending node** (☊). Experience has shown that disturbances occur when the Moon is at its perigee or its nodes and these are unfavourable times for cultivating the soil, sowing and harvesting.

On the calendar (pages 56-79), the times to be avoided are indicated in red. Obviously, the closer we are to the nodes and the perigee, the more harmful the influence. Since this influence develops gradually, we cannot indicate precisely when it begins and ends; to get an approximate idea of the hours to avoid, however, simply check the time that the red zone begins and ends.

Planetary nodes

The planets also have nodes, just like the Moon. We have taken them into account up to the present in the last band of the calendar on pages 56-79. This year we include them in the note pages (81–108) using the node symbols (☋ ☊) followed by the appropriate planet symbols (☿ ♀ ♂ ♃).

Eclipses

An eclipse takes place when the Moon is new or full, when it is aligned with the Sun and the Earth in the plane of the ecliptic (2): when the Moon is new it is an eclipse of the Sun and when the Moon is full it is an eclipse of the Moon. When the line of the lunar nodes intersects the Earth's plane of rotation around the Sun and the Moon's plane of rotation around the Earth (1), an eclipse will take place if the node coincides with the New Moon or the Full Moon.

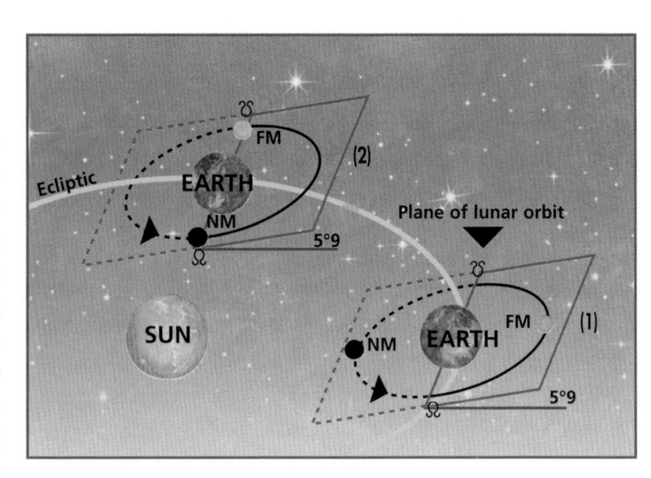

Planes of the EARTH-SUN and MOON-EARTH rotation.

For eclipses in 2011, see the following page.

Eclipses of the Moon in 2011:

- 15 June: total eclipse, reaching its maximum at 22:13 (BST)
- 10 December: total eclipse, reaching its maximum at 15:32 (GMT)

Eclipses of the Sun in 2011:

- 4 January: partial eclipse reaching its maximum at 09:51 (GMT), visible in northwest Africa, western Europe and central Asia.
- 1 June: partial eclipse reaching its maximum at 23:16 (BST), visible in eastern Asia, northern North America, Greenland and Iceland.
- 1 July: partial eclipse reaching its maximum at 10:38 (BST), visible in southern Indian Ocean
- 25 November: partial eclipse reaching its maximum at 07:20 (GMT), visible in southern Africa, Antartica, Tasmania and New Zealand.

Zodiac signs and constellations

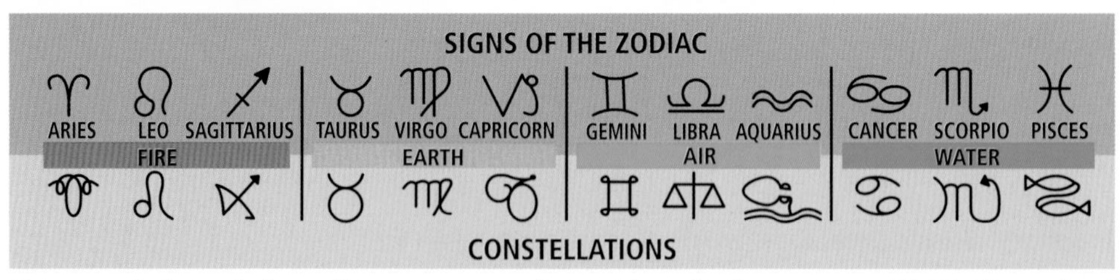

The constellations of the zodiac are groups of stars, named after their various shapes. These constellations are of different sizes. When the Moon passes in front of, or 'through', a constellation, it activates the influences belonging to it and transmits them to us. We owe our understanding of the laws concerning the Moon in the constellations to Rudolf Steiner, the father of anthroposophy. In 1924, he became the founder of biodynamic cultivation and his work has promoted the development of different systems of organic farming.

Centuries ago, the Chaldeans of Mesopotamia imposed a regular structure on the constellations, dividing them into 12 equal parts of 30° in order to mirror the length of their year. These divisions were called signs and the system will be familiar to followers of sun-sign astrology (dates when the sun changes sign and constellation are included, for information only, on the calendar on pages 56-79). The zodiacal year begins with the sign of Aries in spring, at one of the two points in the year (vernal and autumnal) where the ecliptic crosses the celestial equator (imagine the plane of the terrestrial equator extending into space). In biodynamics (see page 19), the same regular structure is imposed on the lunar cycle. An astronomical phenomenon means that the natural rhythm of the constellations does not correspond exactly with the 'imposed' rhythm of the signs, and because the influences of constellations and signs are different (see below), both are used in biodynamics, sometimes in harmony (see page 10).

The influences of signs and constellations are not the same: the signs affect energy by influencing the fundamental qualities (hot, cold, dry, wet), while the influences of the constellations are more physical and directed towards specific zones, such as plant organs.

The Moon in the zodiac signs

The four fundamental qualities (hot, cold, dry, wet) act upon plants largely through the signs of the zodiac.

Hot tends to assist and accelerate plants' metabolism and helps in the exchange of fluids. Crop production is increased. However, if you actively increase heat, plants could dry out, particularly if there is already internal or external dryness. If there is already too much internal or external moisture, the risk of contracting parasitic or viral diseases increases. 'Hot' helps to counterbalance excessive external 'cold'.

Cold tends to slow down the development of the plant, hinder the exchange of fluids and limit metabolic changes. It also improves resistance to heat.

Dry tends to limit the amount of water in the tissues, concentrates the sap and helps to resist external moisture. Too much can cause the plant to mature too rapidly and wither.

Wet governs the amount of water in plants, which can reach more than 99% of plant mass. It stimulates active principles and nourishes the whole plant. Excess fluid can lead to decomposition and rotting.

- These qualities cannot exist in isolation; they combine to create the four elements.
- The intensification and concentration of 'hot' by 'dry' creates **FIRE**. Fire is therefore dry and hot.
- The expansive and vapourizing effects of 'wet' by 'hot' creates **AIR**. Air is therefore hot and wet.
- The condensation of 'wet' by 'cold' creates **WATER**. Water is therefore cold and wet.
- The concentration of 'cold' by 'dry' creates **EARTH**. Earth is therefore cold and dry.

It is not possible, therefore, to apply one quality in isolation. If you want to apply the quality 'dry', you must choose **DRY** and **COLD** (in other words, an **EARTH** sign) or **DRY** and **HOT** (a **FIRE** sign).

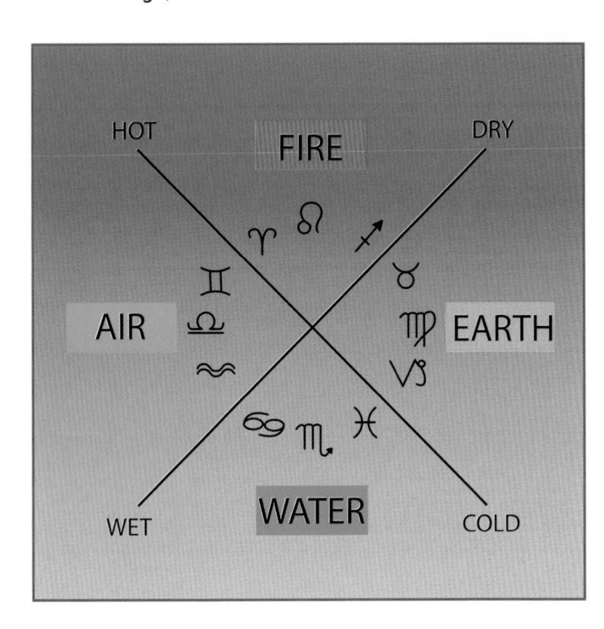

For example, if you want a specifically dry effect, work alternately when the Moon is in an Earth sign and when it is in a Fire sign; the alternate hot and cold qualities cancel each other out and the dry effect of your work will remain.

The Moon in the constellations

When the Moon is in a **Fire constellation** (Aries, Leo, Sagittarius), plant activity is concentrated mainly in the development of **fruits** and **seeds**. It is a good time for growing tomatoes, French beans, peas, apples and cereals of all kinds (and for maintenance and planting seeds, where appropriate).

When the Moon is in an **Air constellation** (Gemini, Libra, Aquarius), the **flowering** part of a plant grows well, so this is a positive phase for cultivating vegetables such as cauliflowers and artichokes, and ornamental flowers.

When the Moon is in a **Water constellation** (Cancer, Scorpio, Pisces), the **leaf** parts of plants grow well. Now is the time to work on salad vegetables, spinach, chard and so on.

When the Moon is in an **Earth constellation** (Taurus, Virgo, Capricorn), **bark** and **roots** develop well, making it an ideal time to concentrate on carrots, potatoes, asparagus, celery, parsnips and other root vegetables.

When signs and constellations are aligned

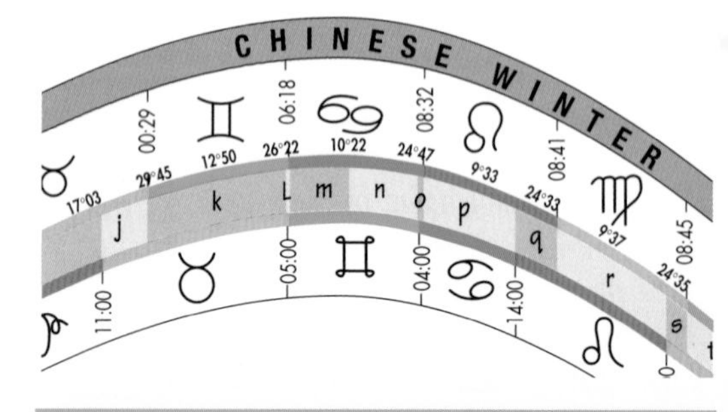

Better results are achieved when certain signs and constellations are aligned. We have concentrated on the effects of these alignments on food and wine crops, but further observations will allow us to focus and expand on the effects on other plants.

These alignments are represented in the bands on the calendar (pages 56-79) by letters of the alphabet. For example, 'p' indicates that the sign of Leo coincides with the constellation of Cancer. Varying shades of grey indicate periods corresponding to a letter and the times of changes are noted within the signs and constellations bands.

Planetary aspects

The planets revolve around the Sun at different distances and speeds. Viewed from the Earth, they move constantly across the sky. If we look at two planets at any given moment, they form an angle of which the Earth is the apex. If we see these two planets in alignment, the angle formed with the Earth is 0°. When this happens, they are in conjunction. When one planet sets and the other rises, it is known as an opposition (180°). When the position of a planet creates an angle of 60°, 90° or 120°, it is in aspect. Each aspect has been given its own colour in order to make location easier.

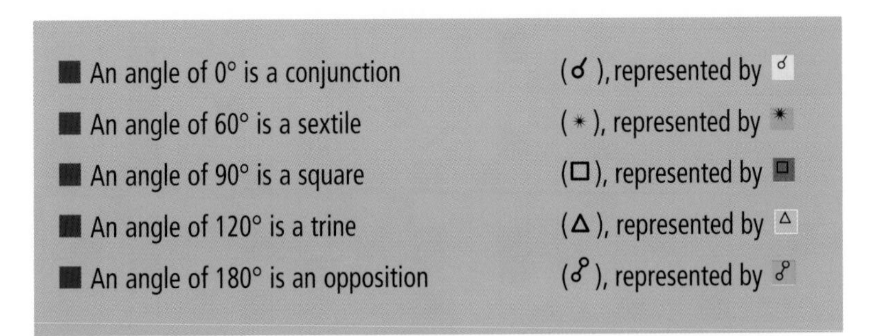

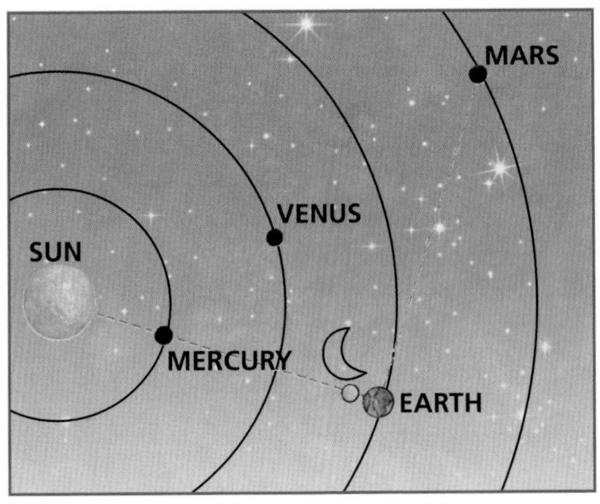

In the illustration, the Moon is in conjunction with Mercury and the Sun (☾☌☿ ☉) and also square to Mars (☾□♂).

Some of these aspects can promote the harmonization of energy, leading to healthier growth. Others can disrupt the plant's energy, causing such problems as slow growth, excessive growth and disease.

In several chapters, we refer to planetary aspects - the days and times of these are indicated in the following pages.

We have observed that some aspects play a major role in various domains, during a time scale varying from a few hours to a whole day. You too can make your own observations and use planetary aspects in other ways than listed here.

For instance, checking in the tables that follow one of the moon's planetary aspects is: January 2 at 14:00 when the Moon is in conjunction to Mercury (☾ ☌ ☿).

JANUARY	1	2	3	4	5	6	7	8	9	10	11	12	13	14	15	16	17	18	19	20	21	22	23	24	25	26	27	28	29	30	31
☉: SUN				☌ 9					✶ 17			12			△ 3				☍ 21					△ 6		13		✶ 23			
☿: MERCURY		☌ 14					✶ 14			8			△ 3					☍ 6				△ 19			1		✶ 11				
♀: VENUS					✶ 12			4		△ 22						☍ 6				△ 21			2		✶ 7					☌ 3	
♂: MARS					☌ 0					✶ 6		23			△ 13					☍ 4				△ 11		17			✶ 2		
♃: JUPITER			2		✶ 11					☌ 11					✶ 11		18		△ 21					☍ 0				△ 9		17	
♄: SATURN		✶ 7		15			△ 1					☍ 2				△ 21			1		✶ 2				☌ 5				✶ 14		23
♅: URANUS			2		✶ 11					☌ 10					✶ 8		15		△ 18				☍ 20					△ 3		10	
♆: NEPTUNE			✶ 2				☌ 21					✶ 22			8		△ 15				☍ 19				△ 22			3		✶ 10	
♇: PLUTO			☌ 18					✶ 14			3		△ 15					☍ 6				△ 9		10		✶ 13					☌ 2
	1	2	3	4	5	6	7	8	9	10	11	12	13	14	15	16	17	18	19	20	21	22	23	24	25	26	27	28	29	30	31

FEBRUARY	1	2	3	4	5	6	7	8	9	10	11	12	13	14	15	16	17	18	19	20	21	22	23	24	25	26	27	28			
☉: SUN			☌ 3					✶ 14			7		△ 20					☍ 9				△ 16		23			✶ 12				
☿: MERCURY	☌ 17						✶ 10			8			△ 2				☍ 23					△ 11		23			✶ 16				
♀: VENUS				✶ 11			6			△ 1					☍ 4				△ 13		17			✶ 0							
♂: MARS			☌ 3					✶ 12			4		△ 17					☍ 4				△ 9		15			✶ 2				
♃: JUPITER		✶ 3					☌ 5					✶ 6		14		△ 18				☍ 18					△ 1		10				
♄: SATURN			△ 9					☍ 9					△ 6		11		✶ 13				☌ 12				✶ 18			3			
♅: URANUS	✶ 20					☌ 19					✶ 19			3		△ 7				☍ 7				△ 11		18					
♆: NEPTUNE				☌ 6					✶ 8		19			△ 3				☍ 7				△ 7		10		✶ 17					
♇: PLUTO				✶ 23			12			△ 1				☍ 18				△ 21		20		✶ 21					☌ 9				
	1	2	3	4	5	6	7	8	9	10	11	12	13	14	15	16	17	18	19	20	21	22	23	24	25	26	27	28			

MARCH	1	2	3	4	5	6	7	8	9	10	11	12	13	14	15	16	17	18	19	20	21	22	23	24	25	26	27	28	29	30	31
☉: SUN				☌ 21						✶ 9		23			△ 10				☍ 18					△ 2		12			✶ 4		
☿: MERCURY					☌ 13						✶ 12			5		△ 16					☍ 1				△ 12			0		✶ 12	
♀: VENUS	☌ 3					✶ 16			11			△ 5					☍ 1				△ 8		13		✶ 23						☌ 9
♂: MARS				☌ 7					✶ 17			8		△ 19					☍ 3				△ 8		15			✶ 4			
♃: JUPITER	✶ 21						☌ 0					✶ 2		10		△ 15				☍ 15				△ 20			5		✶ 17		
♄: SATURN		△ 13					☍ 13					△ 12		18		✶ 21				☌ 19				✶ 22			6		△ 16		
♅: URANUS	✶ 4					☌ 5					✶ 5		15		△ 20				☍ 21				△ 22			3		✶ 14			
♆: NEPTUNE			☌ 15					✶ 16			4		△ 13				☍ 21				△ 19		20			✶ 1				☌ 23	
♇: PLUTO				✶ 7		20			△ 8					☍ 4				△ 9		8		✶ 7				☌ 16					✶ 15
	1	2	3	4	5	6	7	8	9	10	11	12	13	14	15	16	17	18	19	20	21	22	23	24	25	26	27	28	29	30	31

APRIL	1	2	3	4	5	6	7	8	9	10	11	12	13	14	15	16	17	18	19	20	21	22	23	24	25	26	27	28	29	30	31
☉: SUN			☌ 16						✶ 1		13		△ 21					☍ 4				△ 15			4		✶ 20				
☿: MERCURY				☌ 11					✶ 3		8		△ 10				☍ 7				△ 6		11		✶ 20						
♀: VENUS						✶ 0		17			△ 7				☍ 22					△ 6		14			✶ 4					☌ 18	
♂: MARS		☌ 13					✶ 21			10		△ 19					☍ 2				△ 9		18			✶ 7					
♃: JUPITER			☌ 20					✶ 21			6		△ 12				☍ 14				△ 18			1		✶ 12					
♄: SATURN			☍ 16					△ 15		23			✶ 3				☌ 4				✶ 5		10		△ 19					☍ 19	
♅: URANUS		☌ 15					✶ 15			1		△ 8				☍ 11				△ 11		16			✶ 0					☌ 0	
♆: NEPTUNE					✶ 1		13		△ 22					☍ 8				△ 8		9		✶ 12					☌ 7				
♇: PLUTO			3		△ 16					☍ 12				△ 20		20		✶ 19					☌ 1				✶ 21			10	
	1	2	3	4	5	6	7	8	9	10	11	12	13	14	15	16	17	18	19	20	21	22	23	24	25	26	27	28	29	30	31

MAY	1	2	3	4	5	6	7	8	9	10	11	12	13	14	15	16	17	18	19	20	21	22	23	24	25	26	27	28	29	30	31
☉: SUN			☌ 8					✶ 12		22			△ 4				☍ 12					△ 5		20			✶ 14				
☿: MERCURY	☌ 1					✶ 7		19			△ 5				☍ 17					△ 8		23			✶ 19						☌ 17
♀: VENUS						✶ 5		19			△ 5				☍ 17					△ 6		19			✶ 13						☌ 2
♂: MARS	☌ 16					✶ 21			8		△ 15				☍ 23					△ 9		20			✶ 10					☌ 19	
♃: JUPITER	☌ 16					✶ 15			0		△ 6				☍ 10				△ 15		22			✶ 9					☌ 11		
♄: SATURN					△ 17			1		✶ 7				☌ 11				✶ 13		17			△ 0				☍ 23				
♅: URANUS					✶ 0		9		△ 16				☍ 22					△ 0		4		✶ 10					☌ 9				
♆: NEPTUNE		✶ 8		20			△ 5				☍ 16				△ 19		20		✶ 23					☌ 15					✶ 16		
♇: PLUTO		△ 22					☍ 17					△ 3		5		✶ 5				☌ 10					✶ 3		16			△ 4	
	1	2	3	4	5	6	7	8	9	10	11	12	13	14	15	16	17	18	19	20	21	22	23	24	25	26	27	28	29	30	31

JUNE	1	2	3	4	5	6	7	8	9	10	11	12	13	14	15	16	17	18	19	20	21	22	23	24	25	26	27	28	29	30	31
☉: SUN	☌ 22					✶ 20			3		△ 9				☍ 21					△ 20			13			✶ 6					
☿: MERCURY						✶ 4		18			△ 5					☍ 5					△ 18			18			✶ 17				
♀: VENUS					✶ 7		16			△ 0				☍ 14					△ 11			4		✶ 23						☌ 9	
♂: MARS				✶ 19			3		△ 9				☍ 19					△ 9		21			✶ 13					☌ 19			
♃: JUPITER			✶ 9		16		△ 22					☍ 4				△ 11		18			✶ 4					☌ 6					
♄: SATURN	△ 21			5		✶ 11				☌ 17				✶ 21			1		△ 8					☍ 7					△ 5		
♅: URANUS	✶ 9		17		△ 23					☍ 7				△ 11		15		✶ 21					☌ 18					✶ 18			
♆: NEPTUNE	3		△ 11				☍ 22					△ 3		5		✶ 9				☌ 23					✶ 23			10		△ 19	
♇: PLUTO			☍ 22					△ 8		11		✶ 13				☌ 18					✶ 10		22			△ 10					
	1	2	3	4	5	6	7	8	9	10	11	12	13	14	15	16	17	18	19	20	21	22	23	24	25	26	27	28	29	30	31

JULY	1	2	3	4	5	6	7	8	9	10	11	12	13	14	15	16	17	18	19	20	21	22	23	24	25	26	27	28	29	30	31
☉: SUN	☌ 10					✶ 1		7		△ 14					☍ 8					△ 12			6		✶ 23					☌ 20	
☿: MERCURY			☌ 1				✶ 20			4		△ 13					☍ 13					△ 23			14			✶ 2			
♀: VENUS					✶ 4		12		△ 19					☍ 14					△ 20			15			✶ 10					☌ 11	
♂: MARS			✶ 14		20			△ 1				☍ 13					△ 8		22			✶ 14					☌ 18				
♃: JUPITER	✶ 2		8		△ 12				☍ 18					△ 4		11		✶ 21					☌ 22					✶ 18		22	
♄: SATURN	13		✶ 17				☌ 23					✶ 6		11		△ 19					☍ 17					△ 17			0		✶ 4
♅: URANUS	1		△ 7				☍ 13				△ 19		23			✶ 6					☌ 3					✶ 2		10		△ 15	
♆: NEPTUNE					☍ 3				△ 9		12		✶ 16					☌ 7					✶ 7		18			△ 3			
♇: PLUTO	☍ 4				△ 12		15		✶ 18					☌ 2				✶ 17			5		△ 17					☍ 12			
	1	2	3	4	5	6	7	8	9	10	11	12	13	14	15	16	17	18	19	20	21	22	23	24	25	26	27	28	29	30	31

AUGUST	1	2	3	4	5	6	7	8	9	10	11	12	13	14	15	16	17	18	19	20	21	22	23	24	25	26	27	28	29	30	31
☉: SUN				✶ 6		12		△ 20					☍ 20						△ 5		23			✶ 13					☌ 4		
☿: MERCURY	☌ 11				✶ 15		17		△ 20					☍ 6				△ 22			7		✶ 16					☌ 1			
♀: VENUS				✶ 0		7		△ 16					☍ 18						△ 7			2		✶ 18					☌ 10		
♂: MARS	✶ 7		11		△ 15					☍ 6					△ 5		20			✶ 12					☌ 14				✶ 23		
♃: JUPITER		△ 1				☍ 5				△ 15		23			✶ 9					☌ 10					✶ 6		11		△ 12		
♄: SATURN				☌ 8				✶ 15		22			△ 6					☍ 5					△ 6		14		✶ 18				☌ 20
♅: URANUS			☍ 18					△ 0		5		✶ 13					☌ 9					✶ 9		19			△ 0				
♆: NEPTUNE	☍ 10				△ 13		16		✶ 21					☌ 13					✶ 13			1		△ 11				☍ 18			
♇: PLUTO	△ 19		20		✶ 22					☌ 7					✶ 0		11			△ 0				☍ 21					△ 3		3
	1	2	3	4	5	6	7	8	9	10	11	12	13	14	15	16	17	18	19	20	21	22	23	24	25	26	27	28	29	30	31

SEPTEMBER	1	2	3	4	5	6	7	8	9	10	11	12	13	14	15	16	17	18	19	20	21	22	23	24	25	26	27	28	29	30	31
☉: SUN		✶ 12		19			△ 5					☍ 10					△ 22			15			✶ 2				☌ 12				
☿: MERCURY	✶ 4		9		△ 18						☍ 2						△ 0		23			✶ 16					☌ 10				
♀: VENUS		✶ 20			5		△ 17						☍ 3					△ 18			11		✶ 22					☌ 8			
♂: MARS	1		△ 5				☍ 22						△ 0		16			✶ 8					☌ 7				✶ 14		15		
♃: JUPITER		☍ 13				△ 22			6		✶ 16					☌ 16					✶ 14		19		△ 21				☍ 19		
♄: SATURN					✶ 1		8		△ 18					☍ 17					△ 19			5		✶ 10				☌ 11			
♅: URANUS				△ 4		9		✶ 17					☌ 14					✶ 15			1		△ 8				☍ 10				
♆: NEPTUNE	△ 19		21			✶ 2				☌ 19					✶ 18			7		△ 18					☍ 4				△ 3		
♇: PLUTO		✶ 4				☌ 12					✶ 6		18			△ 6					☍ 5				△ 14		14		✶ 13		
	1	2	3	4	5	6	7	8	9	10	11	12	13	14	15	16	17	18	19	20	21	22	23	24	25	26	27	28	29	30	31

OCTOBER	1	2	3	4	5	6	7	8	9	10	11	12	13	14	15	16	17	18	19	20	21	22	23	24	25	26	27	28	29	30	31
☉: SUN	✶ 20			4		△ 17						☍ 3					△ 15			5		✶ 14				☌ 21					✶ 6
☿: MERCURY		✶ 0		13			△ 6					☍ 23						△ 19			9		✶ 18					☌ 3			
♀: VENUS		✶ 19			7		△ 23						☍ 13						△ 2		14		✶ 22					☌ 5			
♂: MARS	△ 18					☍ 12					△ 17			9			✶ 0				☌ 21					✶ 3		4		△ 6	
♃: JUPITER				△ 0		8		✶ 18					☌ 17					✶ 16		23			△ 2				☍ 1				△ 1
♄: SATURN		✶ 14		20			△ 5					☍ 6					△ 8		18			✶ 1				☌ 3				✶ 2	
♅: URANUS	△ 10		13		✶ 20					☌ 18					✶ 19			6		△ 14				☍ 19				△ 18		19	
♆: NEPTUNE	3		✶ 7				☌ 23					✶ 23			12		△ 23					☍ 13				△ 13		13		✶ 14	
♇: PLUTO			☌ 18					✶ 12			0		△ 13					☍ 13					△ 0		1		✶ 0				☌ 2
	1	2	3	4	5	6	7	8	9	10	11	12	13	14	15	16	17	18	19	20	21	22	23	24	25	26	27	28	29	30	31

NOVEMBER	1	2	3	4	5	6	7	8	9	10	11	12	13	14	15	16	17	18	19	20	21	22	23	24	25	26	27	28	29	30	31
☉: SUN		17			△ 8					☍ 20						△ 4		15		✶ 22					☌ 6				✶ 20		
☿: MERCURY	✶ 20			12			△ 7					☍ 21						△ 0		7		✶ 10				☌ 10				✶ 14	
♀: VENUS	✶ 21			13			△ 7					☍ 22						△ 4		13		✶ 19					☌ 4				
♂: MARS				☍ 0					△ 6		21			✶ 10					☌ 5				✶ 12		13		△ 16				
♃: JUPITER		7		✶ 16					☌ 15					✶ 13		22			△ 3				☍ 5				△ 6		10		
♄: SATURN	8		△ 17					☍ 18					△ 19			5		✶ 12				☌ 18				✶ 19		23			
♅: URANUS		✶ 0				☌ 21					✶ 22			9		△ 18					☍ 3				△ 3		4		✶ 8		
♆: NEPTUNE				☌ 4					✶ 4		16			△ 4				☍ 19				△ 23		23			✶ 0				
♇: PLUTO				✶ 18			6		△ 19					☍ 18					△ 9		11		✶ 12				☌ 13				
	1	2	3	4	5	6	7	8	9	10	11	12	13	14	15	16	17	18	19	20	21	22	23	24	25	26	27	28	29	30	31

DECEMBER	1	2	3	4	5	6	7	8	9	10	11	12	13	14	15	16	17	18	19	20	21	22	23	24	25	26	27	28	29	30	31
☉: SUN		10			△ 3					☍ 15					△ 16			1		✶ 7				☌ 18					✶ 14		
☿: MERCURY		18			△ 0				☍ 14					△ 5		12		✶ 17					☌ 3				✶ 19			10	
♀: VENUS		✶ 4		23			△ 19						☍ 4					△ 2		10		✶ 16					☌ 8				
♂: MARS		☍ 10					△ 15			9		✶ 16					☌ 8				✶ 15		18		△ 21					☍ 14	
♃: JUPITER	✶ 17					☌ 17					✶ 15		23			△ 5				☍ 11				△ 14		18			✶ 0		
♄: SATURN	△ 7					☍ 7					△ 7		16		✶ 23					☌ 6				✶ 10		14		△ 20			
♅: URANUS				☌ 3					✶ 4		15		△ 23					☍ 9				△ 13		15		✶ 19					☌ 11
♆: NEPTUNE	☌ 11					✶ 11			0		△ 10					☍ 1				△ 8		10		✶ 12				☌ 22			
♇: PLUTO		✶ 3		15			△ 4					☍ 2				△ 16		20		✶ 22					☌ 2				✶ 13		
	1	2	3	4	5	6	7	8	9	10	11	12	13	14	15	16	17	18	19	20	21	22	23	24	25	26	27	28	29	30	31

The Red Moon

This period begins with the New Moon immediately after Easter and ends at the following New Moon. Since Easter is the Sunday following the first Full Moon of spring, **this year the Red Moon lasts from 3 May to 1 June.**

During this time, young shoots and buds can become damaged by the cold even when the air temperature is above zero. This phenomenon is probably due to the balancing of the temperature in the upper atmosphere with the soil temperature, rather than to the light of the Moon. In fact, when the sky is clear, the temperatures are balanced by radiation, which is greater at soil and plant level. As a result, while the temperature in the air may be just above zero, the temperature of the air surrounding plants can fall below zero, scorching young buds and leaves. This balancing occurs throughout the year but only produces this effect during the period of the Red Moon due to the combination of fragile buds and the low soil temperature.

May Frost

A drop in temperature and the last of the spring frosts usually takes place during May; this normally occurs around **11, 12 and 13 May.**

Indian Summer

Around 11 November, six months after the Blackthorn Winter in May, the exact opposite takes place, and an un-seasonal rise in temperature occurs, known as an Indian Summer.

At both times, it appears as if the previous season is making one last effort to assert itself before the next season takes over.

Tides

You might think that the influence of the tides is restricted to the movement of water, but in fact tidal influence also affects the soil and most plants. Tides are caused by the gravitational attraction of astronomical bodies; their strength depends on the mass of the bodies and the distance between them. As far as the Earth is concerned, the two bodies with the most significant effect are the Moon and the Sun. The Moon's effect is, on average, 2.17 times stronger than that of the Sun. This attraction subtly alters the shape of the Earth, and this phenomenon is most apparent in the movements of the oceans. As centrifugal force counterbalances the strength of gravitational attraction, the Earth's ocean surface which is naturally shaped like an egg (1), is pulled into the shape of a rugby ball (2).

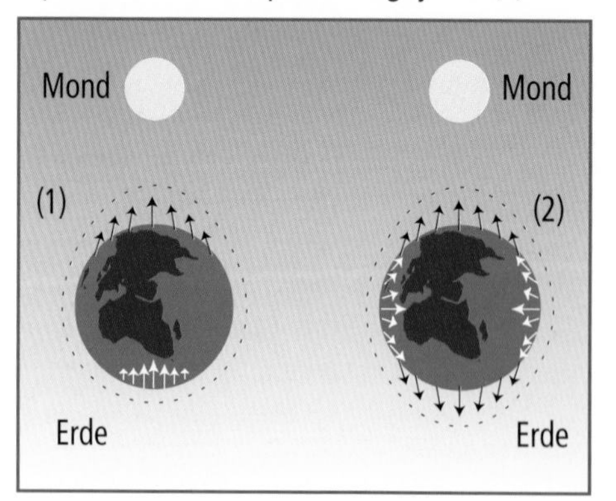

Since the attraction of the Moon is greater, its position determines the tides. The effect of the Sun is limited to reducing or increasing the Moon's gravitational attraction. When the Moon is new or full, the Sun, Moon and Earth are aligned, so the two gravitational pulls combine to cause the highest tides (spring tide). During the first and last quarters, on the other hand, the lunar and solar gravitational attractions oppose one another, so the tides are lower (neap tide).

Atmospheric pressure has a delaying effect on the natural rhythm of the tides, but without this influence the tide would always be at its lowest when the Moon is ascending. During the time the Moon takes to reach its highest point, or meridian (on average 6 hours and 12 minutes), the tide rises (incoming tide) before beginning to recede (ebb tide) until the Moon sets. The tide then rises again to reach the opposite meridian (+180°) before receding once more until the following evening. This double cycle of rising and receding tides takes 24 hours and 50 minutes, the time the Moon takes to return to the same meridian and a little longer than one rotation of the Earth (to compensate for the Moon's displacement during the Earth's rotation).

It is not only the seas that are affected – scientists have shown that the Earth's crust lifts 30 cm (12 in) or more during high tides and it is also known that the tides can affect living creatures. It seems obvious to us that tides have an influence on sap and on crops in general. Although we have not been able to gather together enough evidence to verify the following points, and despite the practical difficulties involved, we consider it useful to take them into account when carrying out certain tasks.

Choose a rising tide for the following:
- sowing, when the Moon is descending to balance energies
- grafting
- cultivating heavy soil
- spreading compost (avoid compacting the compost)
- harvesting, when the Moon is waning

Choose a receding tide for these tasks:
- sowing, when the Moon is ascending to balance energies
- pricking out
- cultivating light soil
- cutting wood
- harvesting, when the Moon is waxing

Tidal influences are not felt immediately, and with plants, the delay is estimated at about an hour. For example, the first effects of a rising tide will be felt about one hour after the Moon rises or sets; the effects will continue until one hour after it reaches the meridian. In the time (up to thirty minutes, in fact) that precedes or follows the change, the effect will be greatly reduced, so avoid these times if you want to get the maximum benefit from your work. The effects of a rising tide are stronger in the morning, while those of a receding tide are stronger in the afternoon. This pattern occurs near the New Moon and the Full Moon; tidal influences are weaker around the 'quarters'.

The times given in the notes pages (81–108) are for GMT standard time (26 October to 29 March) and British Summer Time (30 March to 25 October) – see page 80 for other countries.

The times given for the rising and setting of the Moon are for London. For the times for where you live, see the table, page 80.

On pages 81-108, to make the calculations easier, the time that a rising tide begins to affect plants is marked in green and the time that a receding tide begins to affect plants is marked in black. The times for London are valid for a good part of England. For greater precision, see the time differences marked on the map, right.

TIME DIFFERENCES FROM LONDON

For Ireland add 30 minutes to the times indicated. For example, if you want to carry out a crown graft on 3 May and benefit from the effect of the rising tide, do it between 06:15 and 14:00 if you are in the London area. If you are in Ireland, do it between 06:45 and 14:30.

Chinese seasons

The traditional Chinese year is made up of 12 lunar months (13 months in 7 years out of a 19-year cycle), each month beginning at the New Moon and lasting 29 or 30 days. The Chinese New Year takes place between 21 January and 20 February. Chinese springtime is determined by the Sun and begins halfway between the winter solstice and the vernal (spring) equinox. Between each season comes an 'inter-seasonal' period of 18 days that makes it easier to adapt to the change from one season to the next. According to traditional Chinese culture Qi (pronounced chi) is the energy or 'life force' that is present everywhere in the universe and that flows around us as part of that universe. The seasons influence the nature of the Qi. The five elements (wood, fire, earth, metal, water), and all that is linked to them, form the basis of Chinese culture and medicine (see the table below). The main focus of Chinese medicine is on the prevention of illness by taking into account the rhythm and flow of the Qi – treating the patient at the appropriate time enables treatment to work more effectively.

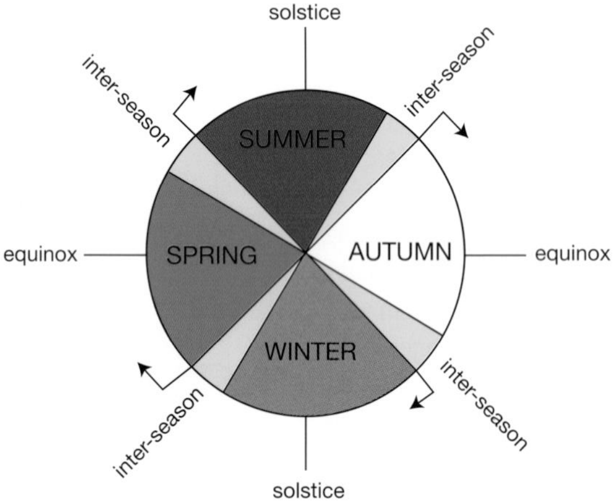

The method used in acupuncture is simple and effective – treatment is given during the inter-seasonal period before the season that governs the part of the body that is affected. For example, for problems involving the element 'water' (such as bladder, kidneys, fatigue and anxiety), treatment is carried out between autumn and winter. By taking into account the patient's Qi, the most effective days for treatment can be selected. As a result, knowledge of the Chinese seasons is extremely useful for practitioners whose work involves energy flow.

ELEMENT	WOOD	FIRE	EARTH	METAL	WATER
SEASON	Spring	Summer	Inter-season	Autumn	Winter
DIRECTION	East	South	Centre	West	North
ORGAN	Liver	Heart	Spleen	Lung	Kidneys
VISCERA	Gall bladder	Small intestine	Stomach	Large intestine	Bladder
TASTE	Acid	Bitter	Sweet	Pungent	Salty
COLOUR	Greeny blue	Red	Yellow	White	Black
ORGAN/SENSE	Eye/sight	Tongue/speech	Mouth/taste	Nose/smell	Ear/hearing
EMOTIONS	Anger	Joy	Thoughtfulness	Sadness	Intense fear
ENERGY	Wind	Heat	Wet	Dry	Cold

Para Magnetism

The Earth is an electromagnetic body, with its own electromagnetic grid pattern. This grid has an effect on all living things and includes telluric lines, underground streams, ley lines and geological fault lines.

For the past fifty years, physicists, doctors and dowsers have demonstrated how a specific location, or soil, can influence the growth or behaviour of plants, trees, animals and people. Dr Hartmann and Dr Curry are both well-known exponents of the theory, having highlighted the existence of electromagnetic fields.

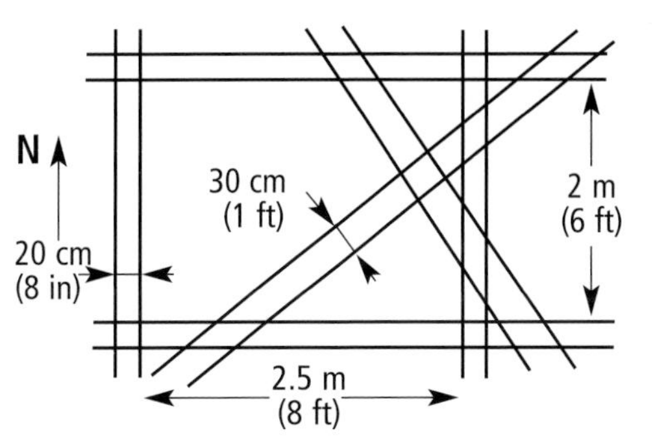

The **Hartmann** Grid consists of naturally occurring charged lines measuring 20 cm (8 in) in width and occurring at intervals of, on average, 2 m (6 ft) running from north to south and 2.5 m (8 ft) from east to west. The grid is like a large-meshed fishing net covering the Earth, and its influence can be felt even at the top of the highest buildings. The **Curry** lines measure 30–40 cm (12–16 in) in width and run diagonally to the Hartmann lines and at irregular intervals.

Many animals avoid the lines, including dogs, cattle, horses, pigs, deer, stags and wild boar. Others are drawn to them, such as cats, ants and swallows.

This sickly tree has suffered as a result of being planted on an intersection of electromagnetic lines (causing the double trunk, canker, stunted appearance, smaller leaves and fruit).

It is advisable not to spend too much time where the lines occur and to take great care over the positioning of furniture and equipment, including beds, workstations and the chair used regularly for watching television. It is important to remember that every living thing positioned directly above the intersection of these lines (the nodes) will be adversely affected at some stage. Pay particular attention to where you plant trees and shrubs and position beehives and animal hutches.

You may already have experienced the effects of electromagnetic fields. If not, it is relatively easy to locate the fields after a few lessons on using a divining rod, a dowsing pendulum, a Hartmann lobe or your own hands.

Hartmann lobe

WORKING WITH THE MOON

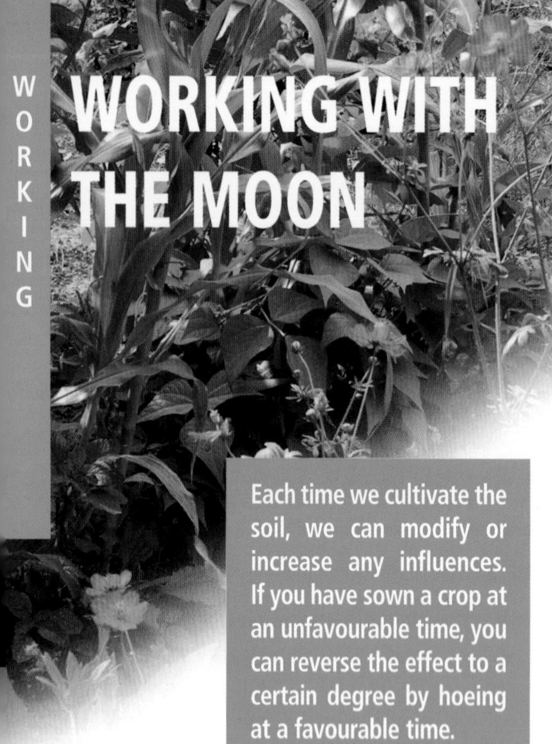

Each time we cultivate the soil, we can modify or increase any influences. If you have sown a crop at an unfavourable time, you can reverse the effect to a certain degree by hoeing at a favourable time.

SOWING: always choose a constellation appropriate to the crop you are sowing (ie. favourable to fruit, root, flower or leaf plants) and preferably sow in the morning. If possible choose one with the maximum number of stars indicated in the lowest band (summarising lunar and planetary influences) on the calendar on pages 56-79.

PLANTING AND PRICKING OUT: choose days when the Moon is descending and also, if possible, when it is opposite a constellation appropriate to the crop that you are growing (ie. favourable to fruit, root, flower or leaf plants) and preferably in the afternoon.

HOEING: for eradicating weeds see page 33.
IN WET WEATHER: if possible, weed in the morning in a Fire or Air sign when the Moon is waning.
IN DRY WEATHER: weed in the evening, if possible, and in an Earth or Water sign when the Moon is waxing.

WATERING: to avoid plants developing shallow roots, instead of watering little and often, water them generously but less frequently. The ideal time is when the Moon is in the descendant and in the constellation of Virgo, Gemini or Libra. Avoid nodes and the superposition of the sign of Leo with the constellation of Leo.

MULCHING: mulch helps to fertilize and protect the soil, and limits the evaporation of moisture and weed germination.
A variety of materials can be used as mulch, such as straw, grass cuttings (a thin layer, fresh or dried), composted bark (very effective, a thickness of 2 or 3 cm/about 1 inch is enough).
It is best to mulch when the Moon is waxing and in an Earth or Air sign.
Aeration of the soil should ideally be carried out when the Moon is ascending; soil decomposes and breaks down more easily (via worms and micro-organisms) when the Moon is descending.

STRAWBERRIES: separate and plant out runners during the constellation of Leo ♌ preferably when the sign of Leo ♌ is in superposition with the constellation of Leo ♌ (q) for example, from July 4 at 07:00 (GMT) to July 5 at 03:16 (GMT).

THE END PRODUCT: always bear in mind the kind of crop you are growing. Take endives as an example – sow and hoe them on dates favourable to 'root' plants in order to help the roots develop well, but, to ensure a good crop, harvest them on dates favourable to 'leaf' plants when the Moon is descending. Their growth should be forced on a good 'leaf' date when the Moon is ascending.

GREEN SALAD: until July, sow green salad in a Water constellation when the Moon is waning in order to prevent these plants from going to seed. During the autumn many forms of energy are in decline. To compensate, sow green salad in a Water constellation, but when the Moon is waxing.

POTATOES: plant potatoes on a day favourable to 'root' plants, but not too close to the perigee. To raise seed potatoes, plant when the Moon is in the sign of Taurus.
To avoid producing green potatoes, earth up when the Moon is in an Earth constellation and when it is waning and also, if possible, with a receding tide.
To remove the eyes (buds) from potatoes that have been lifted and stored so that they will keep longer, choose a time when the Moon is waning and descending, and preferably when it is in the constellation of Virgo.

CROP ROTATION: see page 32.

Seeds

Generally, harvest the seeds in dry weather, when the moon is waxing, avoiding the first and the last day of this phase. Also avoid lunar nodes and times when the Moon is squared with either the Sun or with Saturn.(☾ □ ☉ - ☾ □ ♄).

- **Fruits:** harvest them on a 'fruit' date and wait for a day before extracting the seeds, for instance: tomatoes harvest on August 10 and extract seeds on 11.
- **Roots:** harvest the seeds on a 'root' date.
- **Flowers:** harvest the seeds when the moon is in the constellations Gemini or Libra.
- **Leaves:** harvest when the moon is in an air or earth sign.

Daily rhythms

The Earth's energy patterns vary according to the seasons. During spring, when the Sun is ascending, the Earth 'breathes out', whereas in the autumn, the Sun descends and the Earth 'breathes in'. The same pattern is echoed by the Moon when it is ascending (sometimes called the lunar spring) and when it is descending (also known as the lunar autumn). And energies rise and fall throughout the day too, rising in the **morning** when the Earth 'breathes out' – a good time to **sow** (for example, carrots during the morning of a day favourable to 'root' plants), to **weed** in wet weather, and to **harvest aerial plant parts**. In the **afternoon**, the Earth 'breathes in' so it is a good time to **plant, prick out, plough, weed** in dry weather and **harvest root crops**. The intervening period, from 12:00 to 15:00, is a period of transition and is best avoided.

NEW: Every year, we will focus on a couple of vegetables, one well-known one not so well known that is worth trying.

Potatoes

Climate: Prefers a cool, temperate climate.

Soil: Likes an open, well-drained, sunny spot; dig over the soil, breaking up clods of earth and removing stones. Wait for 3 to 4 years before planting potatoes in the same patch of ground.

Fertilizing: If possible grow a green fertilizer like clover or alfalfa prior to growing potatoes. This must be cut and shredded to the roots (2.5cm, 1 inch deep) and left to dry at least two weeks before planting the potatoes in order not to bury fresh vegetable matter that would attract the click beetle larvae. Dig in plenty of well-rotted manure and compost in the autumn if none has been added for a previous crop (see p.31). Rake in an organic fertilizer about two weeks prior to planting. Potatoes love potash if you have wood ashes, spread about 1kg per 10 m^2 (2lbs per 10 square yards).

beetle larvae

Different types: choose according to need.

- An early one such as Colleen, Accord, Cosmos or Maris Baird
- A waxy one to bake, roast, etc. (Desiree, Robina, Cara, Milva or Valor)
- A floury one to make mashed potatoes, soups and fries (Arran Victory, Home Guard, Maris Piper or Kerr's Pink). It's good to test different types to choose the ones that work best in your soil.

Planting: Choose 'root' dates. Plant every 25 to 30 cm (every foot or so) and space ranks by 45 to 70 cms (every eighteen inches to two feet or so) in order to facilitate ridging up. Cover the seedlings with 5 to 10 cms (2–3") of soil, up to 15 cms of light soil (4"). Use seed potatoes that you have previously chitted, ideally organic. The smaller seed potatoes will yield fewer potatoes but larger ones, the larger seed potatoes will yield more but smaller potatoes.

Plant:

- In March, for an early crop, facing the sun, remember to protect with fleece in case of frost.
- In April, for a seasonal crop – remember to plant on a root day.
- In early May, for a later crop.

Maintenance: Hoe once the leaves appear, then keep ridged (earthed), to protect the leaves from frost, until the ridge is 20 to 25 cms (8–10") in height.

To protect and reinforce the plants, regularly spray (every 2 to 3 weeks):

- Nettle decoction to stimulate growth and prevent disease,
- Tansy decoction to keep away potato beetles,
- Horsetail decoction to avoid mildew.

Horsetail and nettle are usually sufficient to prevent mildew attacks. Only use Bordeaux mixture if climate conditions are really adverse (as systematic use is bad for the soil through copper accumulation). Rock powder (basalt) dusted dry on the leaves (renew after heavy rain) is also a great way to avoid disease and insects, particularly aphids. For potato beetles, there are many organic insecticides based on rotenone or pyrethrum (page 30). Alternatively it is easy to pick off the adults on a regular basis and/or squash eggs under leaves.

Harvest (page 27): 2 to 4 months after planting, depending on the type. Most potatoes are harvested between the end of August and the end of September. On main crop potatoes when the leaves start to dry, cut them off, let the potatoes mature for at least two weeks before harvesting them in dry weather when the moon is descending. Let them dry on the spot or somewhere protected from bad weather, before storing them in a cool store, with little ventilation, keep dark and protected from frost. Light leads to the potatoes getting green, producing solanine, a poison. Solanine, on the other hand, is a positive factor for those potatoes that you mean to plant next season so these can remain for several weeks on the ground whatever the weather except frost.

Tomatillos

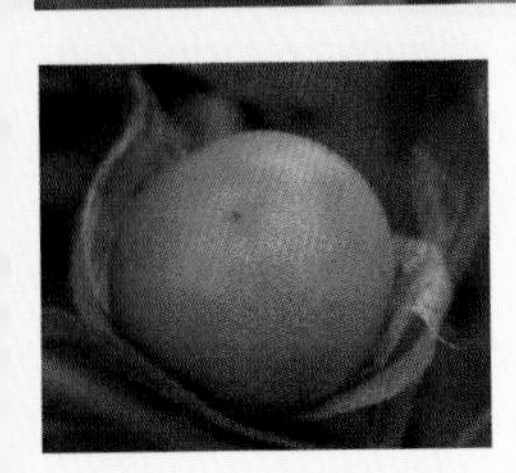

Tomatillos are relatives of the tomato plant and can be grown in exactly the same way. They are used is salsa and the ripe ones can also be eaten raw. They look like green tomatoes with a papery outer casing. There are also yellow and purple streaked varieties.

Climate: Prefers a warm, temperate climate.

Soil: Likes a sheltered, well-drained, sunny spot; dig over the soil, breaking up clods of earth and removing stones. Wait for 3 to 4 years before planting tomatillos in the same patch of ground. As they are members of the Solanaceae family do not grow on ground that grew potatoes in the previous year or near to where you are growing potatoes.

Fertilizing: Dig in plenty of compost in the autumn if none has been added for a previous crop (see p.31). Rake in an organic fertilizer about two weeks prior to planting. Tomatillos love potash if you have wood ashes, spread about 1kg per 10 m2 (2lbs per 10 square yards).

Sowing: Timing is very important when sowing Tomatillo seed. The seed should be sown 7 – 8 weeks before the last frost is expected. Sow on a 'fruit' date. Take a tray of moist organic seed compost, lightly press the surface level and make holes in the surface about 3.5-5cm (1-2in) apart and 0.5cm (¼in) deep. Place one seed in each hole; cover the seed with compost to surface level and gently water with a fine spray.

If a propagator is available set the thermostat at 20°C (68°F) and cover the propagator to exclude light.

Once the seeds have germinated, grow the seedlings on in maximum light.

Pot them on into 13cm (5in) pots when they are 5 – 6cm (approx 2in) tall.

Plant:

- In a greenhouse or polytunnel border, or if space is tight in 30cm (12in) pots filled with good organic compost. Water regularly and ensure that there is plenty of air circulating in the greenhouse or polytunnel to deter mildew.
- Alternatively in warmer areas harden of the seedlings and plant in your prepared bed once all risk of frost is past. Plant 25-30cm (10-12in) apart in either a single or a double staggered row. Stake the tomatillos or they will sprawl. Keep watered in dry weather.

Maintenance: Hoe regularly on a 'fruit' day, both in the greenhouse and in the open ground and weed around tomatillos grown in containers.

To protect and reinforce the plants, regularly spray (every 2 to 3 weeks):

- Nettle decoction to stimulate growth and prevent disease,
- Horsetail decoction to avoid mildew.

Horsetail and nettle are usually sufficient to prevent mildew attacks. Only use Bordeaux mixture if climate conditions are really adverse (as systematic use is bad for the soil through copper accumulation).

Harvest (page 27): the Tomatillos are ready to harvest when the fruits are firm and husks are papery and straw-colored. Usually the husks will break open when they are ripe. If they don't, simply test them with a gentle squeeze to check ripeness. General around eight weeks from planting out but can take longer in cooler climates.

Horticulture

BULBS: plant bulbs when the Moon is in the constellation of Libra (♎).

REPOTTING: repot plants when the Moon is descending.

- Choose a date favourable to 'leaf' plants for those grown for their foliage (♋ or ♏).
- For flowering plants, choose a date favourable to 'flower' plants (♊ or ♎).

FERTILIZING: add fertilizer when the Moon is descending and if possible, when it is waxing, avoiding Fire signs.

GRAFTING ROSES: shield budding can be carried out during any month of the year on a date favourable to 'flower' plants when the Moon is ascending. Graft as low down the stem as possible towards the base to prevent suckers from forming.

PRUNING ROSES: choose a 'flower' date when the Moon is descending. All roses should be pruned when they are dormant or semi-dormant, ie. between autumn and when the buds are just beginning to break out in the spring, but avoiding frosty periods. Prune to an outward facing bud, removing dead, diseased and dying shoots and any crossing growth. Pruning will also help to prevent damage to tall stems caused by wind rock. Most modern roses flower on new or the current season's growth and so should normally be pruned fairly hard to stimulate new growth and to encourage flowering. Shrub and old garden roses should be pruned lightly.

- **Cluster-flowering bush roses (Floribunda):** cut back main stems to 30–40 cm (12–16 in) from the ground or, for taller cultivars, to about one third of their length.
- **Large-flowering bush roses (Hybrid Tea):** cut back to 20–25 cm (8–10 in) from the ground, or slightly less severely in very mild areas.
- **Climbing and rambling roses:** climbers should not normally be pruned in their first two years except to remove dead or diseased stems. In subsequent years shorten side shoots and leave main shoots unpruned unless they are too long. Shorten the side shoots of ramblers in the first two years to a vigorous shoot. In subsequent years cut back old or diseased stems to ground level to maintain a good framework and allow air to circulate freely.

- **Shrub and old garden roses:** prune only lightly – these flower on wood that is two or more years old.

PRUNING FLOWERING SHRUBS: remove dead or diseased stems and any suckers at the base of the plant, retaining the strongest stems. Prune above a lateral bud on a date favourable to 'flower' plants with a descending Moon.

- **flowering shrubs (old wood):** prune as soon as flowering is over (eg. forsythia).
- **flowering shrubs (new wood):** prune quite hard when the worst of the cold weather is over at the beginning of spring (eg. buddleia).

Cultivating trees and shrubs

PLANTING

The following rules are important for planting a tree:

Location: avoid very windy positions and see Geobiology information (page 19).

Soil: ensure it is well balanced — for instance, if it is too acidic, add lime-rich material such as spent mushroom compost, if the soil is too heavy, dig in some sharp sand.

Time of year: avoid heavy frosts. It is preferable to plant trees in the autumn, provided that the following winter is not likely to be too hard. You can also plant in spring but you will need to water the young tress more regularly during the following summer. Whichever season trees are planted in they need a lot of attention during their first few years before they become established (watering, aerating the soil, adding compost).

Planting: it is essential to avoid the lunar nodes (do not plant for two days after an eclipse) as well as times when the Moon is square to the Sun, Mars or Saturn (☾ □ ☉ ♂ ♄) – see pages 12–14.

- **Flowering and fruit-bearing trees and bushes:** plant when the Moon is descending, if possible when it is also waxing and when the tide is receding. If the Moon is waning, choose a time that corresponds with a rising tide.
- **Deciduous trees and shrubs:** mainly as for flowering and fruit-bearing trees above, except for trees that grow quickly and those that tend to 'rush' (poplars, maples), when it is better to plant with the Moon ascending and waning and with a receding tide. This configuration produces softer wood but it will encourage this type of tree to grow more quickly and with fewer knots.
- **Conifers:** plant when the Moon is in the descendant and if possible waning. This helps conifers such as firs to develop deeper roots.

PRUNING

As a general rule, prune when the Moon is descending, although an exception can be made when pruning a young tree since the ascending Moon will encourage new wood to grow more quickly. Avoid lunar nodes, when the Moon is square to the Sun, Mars or Saturn (☾ □ ☉ ♂ ♄) – see pages 12–14 – as well as the perigee, the equinox and periods of severe cold.

- **Fruit trees and bushes:** prune when the Moon is descending, and if possible when it is also waning and when there is a rising tide.

- **Flowering shrubs:** prune when the Moon is descending, and if possible when it is also waxing and when there is a receding tide.

- **Conifers, deciduous trees and hedges:** prune when the Moon is descending, if possible when it is waxing and with a rising tide. However, prune cypress trees, pyracanthas and trees and hedges that have reached their desired height when the Moon is waning rather than waxing.

GRAFTING

Choose an ascending Moon, on a date favourable to 'fruit' plants preferably or on a date favourable to 'flower' plants if this is not possible. Remove and heel in the grafts when the Moon is ascending (in December/January in the UK).

- **Crown graft:** this form of graft is preferable to cleft grafting, which often causes canker and fruit scab. Wait until the rootstock begins to flower so that there is plenty of sap rising and graft on a date favourable to 'fruit' plants when the Moon is ascending, for example 3 May.

- **Shield budding:** this should be carried out in August on a 'fruit' day with a rising moon when the sap is rising for the second time, for example 11 August. To ensure plenty of rising sap, water the rootstock during the fortnight before the grafting. Make the 'shield' from the thin bark of the rootstock, with just a little of the wood attached – by leaving a strip of wood, you prevent the grafted bud from falling out.

PROPAGATING WITH CUTTINGS

Ideally, cuttings should be taken just as the Moon is coming to the end of its ascension and should be planted as it begins to descend.

- **Cuttings from bushes:** (such as gooseberries, blackcurrants, etc.) in autumn, on the last day of the ascending Moon, take cuttings of around 10–15 cm (4–6 in) in length from the current year's growth. Gather the cuttings together in bunches of ten or so, and gently hold the stems together with a rubber band, place them in a plastic bag and leave them for a day in the refrigerator. When the Moon begins its descent, plant the cuttings out at an angle of about 45°, leaving a little less than half the cutting above ground. Planting at this angle helps promote their growth. In spring, as soon as the first leaves appear and when the Moon is descending, gently lift the cuttings and place each in a pot full of enriched peat. A few weeks later, when the roots have developed, plant them out or pot them on in soil-based compost. Propagating cuttings in bunches has been seen to increase hormonal activity, resulting in improved growth.

Harvesting

Various factors influence the quality and preservation of fruit and vegetables after harvest. As a general rule, choose an ascending Moon for harvesting plant parts that grow above ground and a descending Moon for parts that grow below ground, but avoid the perigee, lunar nodes and stormy weather. Fruit and vegetables that do not store well will last much longer if you avoid harvesting during Water signs and constellations (Cancer, Scorpio and Pisces). On the harvesting band on the calendar on pages 56-79, look for the days marked with stars – green stars for aerial parts and yellow for roots.

Other factors to take into account:

- Fruit picked when the Moon is ascending will stay juicy for longer and will have more energy-giving properties.
- The ascending Moon will also help plants to develop and ripen.
- The descending Moon promotes the preservation of certain properties in harvested crops.

As there are many different factors that need to be taken into account we have produced a quick-reference summary:

- Harvest fruit that should be allowed to ripen as slowly as possible (strawberries, raspberries, cherries, apricots, peaches, plums, etc.) when the Moon is waxing and descending (if need be, when it is ascending and waning) and if possible on days with one of the following planetary aspects: (☾ ☌ ✶ △ ♀ - ☾ ☌ ☿ - ☾ ☌ ☍ ♉) see pages 12–14.
- For squashes and varieties of pumpkin that do not keep well, choose a waxing (but not ascending) Moon during the following periods: a - b - i - j - k - L - q - r - s - t - z - (see the calendar on pages 56-79) and if possible pick them during one of the following aspects (☾ ☌ ✶ △ ♀ - ☾ △ ☉) see pages 12–14. For example, **AUGUST:** 1*- 2*- 3*- 4*- 5am*- 29**- 30*- 31* **SEPTEMBER:** 1*- 27pm*- 28am**- 28pm*. (am = morning, pm = afternoon).
- When harvesting root crops, follow the guidelines on the calendar (pages 56-79) and try to combine the descending with the waning Moon, except in the case of garlic, onions and shallots, when you should choose the waxing Moon.

We believe that the tides also affect the harvest (see page 16). When the Moon is waxing choose a receding tide and when it is waning choose a rising tide. The best time to harvest crops is when these periods occur in the early morning with dew on the ground, as this gives positive and magnetic energy to the harvested crops.

Plants diseases

On the whole, diseases can be avoided or at least contained through a good soil balance and intense biological activity (microbes and roots). It is essential to stimulate and restructure the soil with the help of living matter. For instance, green fertilizer is one of the best way to obtain such a result. Using 'young' compost can also generate good results in the soil, except for vegetables which require a 'mature' compost.

In most cases, this intensification of soil life will give enough energy to the plants to avoid disease. Plant decoctions also act as a preventative against diseases (see next chapter). Some other products also have a healthy effect on the soil and help prevent diseases. Here are some of the ones we have been testing with success against oidium and mildew.

CHARCOAL

Charcoal reinforces the soil and has a purifying and cleansing effect that prevents cryptogenic diseases in a localized area. It is particularly efficient at preventing the dispersion of spores and seeds. It regulates the soil by assisting in the storage of nutrients. Use natural, pulverized, charcoal, mixed with the soil prior to sowing the seeds, 100 to 150 g/m2. Ideally this should happen during an ascending and waning moon.

SOOT

Soot resulting from burning wood also helps to prevent cryptogenic diseases as it protects the living organisms in the soil. However do not apply if you have already used charcoal, or if the soil is very acid! Mix 200 to 300 g of dry (not wet) soot with 10 litres of water, ideally during a waxing moon (a, b, c, d, i, j, k, l periods), avoiding lunar nodes and Moon-Saturn squares (☾ ◻ ♄). Do not use metallic containers, as these can taint the mixture, instead use wood or sandstone. Let the mixture steep for several days stirring every evening; spray on the soil, renew monthly.

BIRCARBONATE OF SODA

Bicarbonate of Soda has the ability to stimulate and re-balance chalky soils, resulting in a decrease in diseases.
Spray onto the soil at dew time a few weeks before seeding, using between 20 to 50 g by litre of water. 5 to 10 g per litre can be used directly on plants in the same way as grapefruit seed extract (see below).

GRAPEFRUIT SEED EXTRACT

(available commercially)

Grapefruit seed extract reinforces and purifies plants, acting as a disinfectant and fungicide. Spray on plants at the dates shown in the Fungicide tables, using 35 to 40 drops per litre of water, or even 50 drops to avoid mildew. Renew every 10 to 15 days, and after rain.

Plant-based decoctions

A plant-based decoction used as a liquid fertilizer provides a natural way of feeding plants and preventing certain diseases. Those made from stinging nettle or comfrey are best known, but other plants can be used such as horsetail, camomile, marigold, achillea or dandelion, either individually or in combination.

Whichever plant you choose, the process for **making liquid fertilizer** remains the same. Cut up the fresh plants into small pieces and soak them in water (use 1 kg/2 lb of plants per 10 litres/15 pints of water) and leave it to steep for between one and four weeks – use a wooden or stoneware container, or even plastic if necessary, never use a metal container. Stir the liquid from time to time to help it ferment; then filter out the plant pulp. Use the resulting liquid fertilizer diluted in the proportion of 1:10–20, as appropriate.

WHEN TO MAKE A PLANT DECOCTION

- **To protect against fungal disease:** begin making the decoction when the Moon is waxing and, if possible, when it is also descending. The Moon in conjunction, sextile and trine with the Sun (☾ ☌ ✶ △ ☉) as well as superposition (q), will improve its effectiveness even more. Favourable dates are: MARCH: 15 - 17 - APRIL: 13pm - MAY: 8 - 11 before 15:00 - 13am - JUNE: 6pm - 7 - 11 - JULY: 1pm - 4 - 10 - 31 - AUGUST: 1am - 4am - 8pm - 29 - SEPTEMBER: 2 - 27pm - OCTOBER: 1pm. If you also take into account the influence of the rising tide on these dates (see p16), the result will be even better (am = morning, pm = afternoon).
- **To use as a plant feed:** begin making the decoction when the Moon is waxing, and if possible when it is also ascending. The Moon in conjunction, sextile and trine with the Sun (☾ ☌ ✶ △ ☉) will strengthen its influence.

Stinging nettles

Very favourable dates are: MARCH: 5am - 10 - APRIL: 3 after 16:00 - MAY: 3 - JUNE: 2 before 10:00 - SEPTEMBER: 7am - OCTOBER: 6. (am = morning, pm = afternoon). Make the most of the receding tide on these dates to improve the result even more. Liquid fertilizer made from plants should preferably be used when the Moon is descending and, if possible, when it is also on the wane.

DECOCTION OF STINGING NETTLE

- To use as a fertilizer follow the instructions top left, leave the nettles to steep in water for three weeks and then dilute in the proportion of 1:10 if it is to be spread on the soil, but in at least 1:20 if it is to be sprayed on leaves.
- To prevent fungal diseases, let the mixture steep for a week and then dilute it in the proportion of 1:15–20. The fresh plant can also be used – for example, when potting up tomatoes, put a handful of nettles in the planting hole.

Comfrey

DECOCTION OF COMFREY

A decoction made from comfrey helps combat tomato whitefly, but comfrey really comes into its own as a fertilizer. Fresh leaves can be laid on the soil surface, wilted leaves can be dug into the soil, or it can be used in a liquid fertilizer. To make the liquid feed, soak the leaves in water (see top left this page) for two to four weeks and then dilute in the proportion of 1:10 if applying it directly to the soil or 1:20 if spraying it on leaves.

LIQUID FERTILIZER MADE FROM COMPOST

This is made by mixing one part well-rotted compost to ten parts water and leaving it to steep for one to two weeks. It is effective against cryptogenic (fungal) diseases when sprayed on the leaves of a plant and works on two fronts: it reinforces the plant's immune defences as well as destroying certain fungal spores.

Fungicides and insecticides

We recommend the use of natural products, both for the sake of the environment and to produce a better quality of crop. You can also use liquid fertilizers to prevent cryptogenic diseases such as those caused by fungi (see below and page 27).

A decoction of horsetail is very good for controlling fungal disease. Use about 50 g (2 oz) of horsetail per litre of water (1½ pints) and boil it for 20 minutes. Then dilute with water in the proportion of 1:5 and spray on the leaves or directly on the soil, several times between spring and autumn. If you use sulphur or Bordeaux mixture it is also possible to add horsetail decoction, or liquid fertilizer made from stinging nettles or compost in order to increase the efficiency and reduce the dosage.

Insecticides: a decoction of comfrey can be used to treat tomato whitefly and a decoction of many other plants can be used to produce natural insecticides, such as wormwood, pyrethrum, English ivy, etc. The table below indicates those dates that are the most favourable for applying insecticides and fungicides. The greater the number of stars (★) the more effective will be the application.

- favourable for insecticides
- favourable for fungicides

INSECTICIDES (spread early morning) and **FUNGICIDES** (spread evenings)

	1	2	3	4	5	6	7	8	9	10	11	12	13	14	15	16	17	18	19	20	21	22	23	24	25	26	27	28	29	30	31
JANUARY																								★★	★	★	★				
FEBRUARY																				★	★★	★★	★								
MARCH				★★																★★★	★	★	★★								
APRIL																		★	★												
MAY																													★	★	★
JUNE	★★																									★	★	★★			
JULY	★★																★						★	★	★			★	★★		
AUGUST															★★					★	★	★★		★	★★	★					
SEPTEMBER																★	★	★★		★	★	★★					★				
OCTOBER													★★	★	★★		★	★	★★					★	★						
NOVEMBER											★		★	★	★	★★					★	★★	★	★							
DECEMBER											★	★	★					★	★★	★	★			★							
	1	2	3	4	5	6	7	8	9	10	11	12	13	14	15	16	17	18	19	20	21	22	23	24	25	26	27	28	29	30	31

Compost

Composting is the process by which animal or plant waste decomposes and breaks down into a substance that is both nourishing and easily assimilated by the soil's micro-organisms. A combination of plant waste and animal manure is normally required in order to ensure the constituents of the compost are in the correct proportions.

The composting process can normally be activated by adding stinging nettles (with their seeds heads removed) and comfrey, either as plants or in the form of a liquid. Good compost should be well oxygenated and loosely packed, not compacted. When it is ready to use, it should be light, rich and have a pleasant smell of humus, so it is important to make it at just the right moment. We have summarised all the different factors that will help you to produce good compost in the table below, including certain Moon/Pluto aspects.

COMPOST

	1	2	3	4	5	6	7	8	9	10	11	12	13	14	15	16	17	18	19	20	21	22	23	24	25	26	27	28	29	30	31
JANUARY					★	★		★★★	★						★	★										★★					
FEBRUARY			★	★ / ★★	★						★	★										★ / ★★									
MARCH				★★ / ★	★						★	★		★								★★ / ★									★★
APRIL							★	★		★ / ★								★ / ★★									★ / ★★				
MAY				★	★		★ / ★									★ / ★★ / ★									★						
JUNE		★	★									★★ / ★★									★★										
JULY									★ / ★★ / ★★									★★										★			
AUGUST					★ / ★							★	★																		
SEPTEMBER		★ / ★						★	★	★	★★★ / ★★	★																	★★ / ★★		
OCTOBER					★	★	★	★★★	★									★													
NOVEMBER		★	★	★★ / ★★★	★																		★★						★	★	
DECEMBER	★	★★ / ★						★	★											★							★	★	★★★	★	
	1	2	3	4	5	6	7	8	9	10	11	12	13	14	15	16	17	18	19	20	21	22	23	24	25	26	27	28	29	30	31

(★) the more stars, the more favourable the day

- a good time to add to or turn compost heaps that generate a high internal temperature (the majority of compost heaps)
- choose these dates for working on slower acting compost heaps that generate less heat (often preferred by farmers and gardeners who practise biodynamics)
- recommended for surface composting with non-rotted manure (broken down by worms);
- a vertical line dividing a date box indicates morning (left) or afternoon (right)
- a date box divided by a horizontal line indicates it is favourable for a certain type of composting (green for high temperature compost and orange for surface composting)

MANURE: spread manure when the Moon is descending, if possible on days marked orange (see the table above).

Crop rotation

Each plant has its own nutritional needs and different plants draw nourishment from different layers of the soil. As a result, growing the same plant in the same place can lead to an imbalance of soil constituents and even the depletion of some, along with an increase in parasites, weeds and diseases. In most instances, it is a good idea to wait for three to five years before growing the same crop in the same place again. You can reduce this period, or even ignore it entirely, for vegetables such as spinach and lettuce, which have a short growth cycle. On the other hand, extend the period for crops such as strawberries and asparagus where the plant itself remains in the same position for several years before being replaced, with an annual harvest of just its fruit. Tomatoes are an exception, as they can be grown in the same position for several years in succession.

There are three key rules of crop rotation:

THE FIRST RULE involves rotating plants according to whether they are grown for their roots, leaves, flowers or fruits. On a single plot of land, grow 'fruit' plants the first year (nitrogen-fixing), then 'flower' plants, then 'leaf' (nitrogen-hungry), and finish with 'root' plants. So, for example, you could plant squash, courgettes, maize and French beans, followed by cauliflowers, broccoli or green manure (phacelia, lupins, clover) which would not be cut until the flowering stage; then cabbages, spinach, lettuces and leeks, and, finally, potatoes, celery, beetroots, onions and carrots. You could achieve several rotations in the same year – for instance, green manure could precede or follow a crop. In addition to the rotation of the different crops, it is also important to bear in mind which plants grow well together (see page 110).

THE SECOND RULE is not to grow two plants of the same botanical family one after the other. The main botanical families are as follows:

- **Chenopodiaceae:** beetroot, spinach, Swiss chard.
- **Compositae:** artichoke, cardoon, chicory, endive, lettuce, dandelion, black salsify, Jerusalem artichoke.
- **Cruciferae:** all types of cabbage, cress, mustard, turnip, radish.
- **Cucurbitaceae:** cucumber, marrow, courgette, melon, pumpkin.
- **Leguminosae:** broad bean, French bean, lentil, alfalfa, pea, clover.
- **Liliaceae:** garlic, asparagus, chive, shallot, onion, leek.
- **Umbelliferae:** carrot, celery, chervil, parsnip, parsley.
- **Solanaceae:** aubergine, capsicum, pepper, potato, tomato.

THE THIRD RULE concerns the amount of manure a crop needs. It is good practice to follow a crop that requires large amounts of manure with one that requires little or none.

- **Crops requiring large amounts of manure:** aubergine, celery, cabbage, spinach, fennel, maize, leek, pepper, potato, tomato.
- **Crops requiring little:** garlic, chervil, cress, shallot, broad bean, French bean, corn salad (lamb's lettuce), onion, radish, lettuce.

Other crops fall between the two extremes, requiring an average amount.

Agriculture

CULTIVATING THE SOIL: the work of ploughing, planting, pricking out and spreading compost or manure is best carried out when the Moon is descending. If possible, combine the descending Moon with the waxing Moon when working on light, sandy soil, especially when the weather is dry. In the case of heavy clay soil, if possible, combine the descending Moon with the waning Moon, particularly during wet weather.

ERADICATING WEEDS

- Eradication will be most effective during the days following the full moon, or in the middle of the day – with the best results obtained by combining both together.
- Dig over the soil when the Moon is in the constellation of Leo, which favours germination (marked in ■ in the following table). This will encourage the majority of any weed seeds present in the soil to germinate. Then dig over the soil again to remove them.
- It is possible to contain germination by digging over the soil when the moon is in the constellations of Virgo, Libra and Capricorn (marked in yellow □ in the following table).
- **Thistles:** remove thistles when the soil is dry to help prevent them from seeding too easily. They are difficult to eradicate, but for the best results, work in the middle of the day on the dates marked in purple ■ or blue □ on the chart below. These dates are also the best for dealing with convolvulus.
- **Dock weed:** dig up dock weed on dates marked in purple ■ or orange □.

Choose dates marked with stars (⋆).
If you have no option other than to dig the soil on dates when there are nodes, dig it over again a few days later to help reinstate positive influences.

If only a part of the day is favourable, the time is indicated by a vertical line (the left section for the morning, the right for the afternoon – see above and left for the key to the colours) ◫. On days that are divided horizontally, two favourable aspects are recommended, as indicated by the two colours ⊟.
For instance June 9 ⊞ is favourable until noon to help germination, and then tends to limit it. The beginning of the morning is favourable for pulling out thistles and docks.

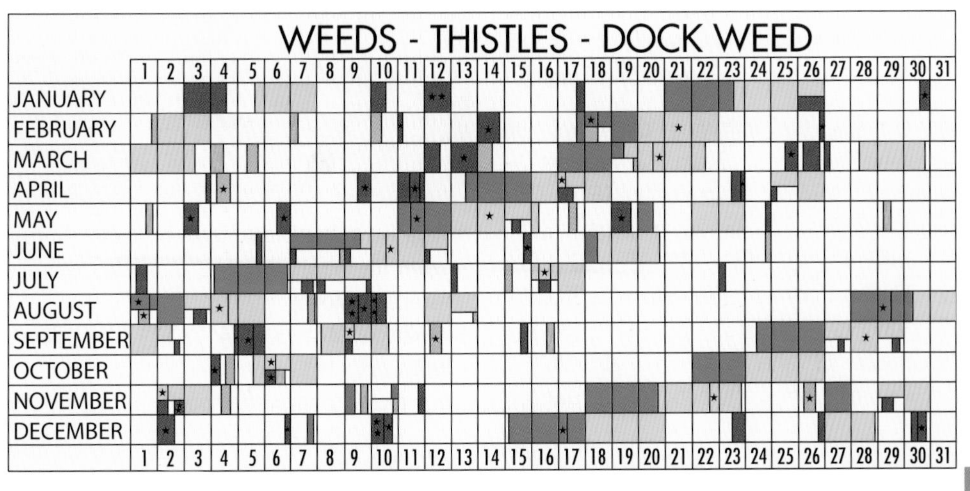

W O R K I N G

CLEARING LAND: to clear land of plants such as thorns and brambles, dig them up or cut them down at the perigee and if possible when the waning Moon is close to becoming new (for instance: SEPTEMBER : 27 - OCTOBER : 26** - NOVEMBER : 24.

MANURE: manure is absorbed by the soil more effectively when the Moon is descending and, if possible, waxing, but avoid spreading manure during Fire signs.

- **Seaweed and natural phosphates (such as in bone meal)**: see 'Manure' left.
- **Leaf mould (minerals):** spread leaf compost in the morning (but in the evening if the forecast is dry and hot), with an ascending and if possible a waning Moon. Avoid Fire signs.
- **Pig manure:** to avoid burning meadow grass, spread chicken manure when the Moon is ascending and, if possible, waxing. Water and Air signs are also preferable.
- **Cattle slurry:** spread cattle slurry when the Moon is waning and avoid Fire signs.

Hay

- **Quality:** hay will be of better quality if it is cut when the Moon is ascending and if possible also waning. The aspects Moon sextile or trine Sun (☾ ✶ △ ☉) will improve the quality still further.

For example, the following dates will be favourable: MAY: 1 - 2* - 20am* - 21* - 22am** - 22pm* - 23* - 24am* - 24pm before 16:00 - 25 - 26 before 08:00 - 27*- 28 - 29am - 29pm* - 30* - 31*- JUNE: 1* - 16* - 17* - 18* - 19* - 20* - 21 - 22 - 23 after 15:00 - 24 - 25 - 26am**- 26pm* - 27* - 28* - 29* - JULY: 15* - 16* - 17* - 18 - 19 - 20 - 21 - 22 - 23* after 08:00 - 24* - 25* - 26*.
(am = morning, pm = afternoon) (in the UK).

When cutting hay, always avoid times that are close to lunar nodes, the perigee (including about 6 hours before and after) and the following aspects: Moon square Sun, Mars and Saturn (☾ □ ☉ ♂ ♄) (about 4 hours before and 2 hours after).

- **Regrowth:** to promote rapid regrowth, cut hay when the waxing Moon is associated with an ascending Moon.

For example: MAY : 3** - 4** - 5* - JUNE : 2am* - JULY : 13* - 14*.
(am = morning, pm = afternoon).

Cereal crops

Sow cereal crops on a day favourable to 'fruit' plants, but if this is not possible, choose days favourable to 'flower' and if not 'root' plants. Days favourable to 'leaf' plants are the least beneficial. Also take the following information into consideration to improve results even further. It is organized according to the type of cereal crop.

OATS, WHEAT, BARLEY, RYE

- **Sow:** preferably when the Moon is ascending and if possible waning, with a receding tide.
- **Harvest:** on days favourable to 'root' or 'flower' plants with the Moon waxing and ascending (on dates favourable to 'fruit' plants there is more dust in the air and the risk of weevils is greater, on 'leaf' dates there is a greater risk of fermentation).
  Examples of favourable dates are: JUNE : 18pm - 19 - 20 - 21 - 22 - 28 - 29 - JULY : 16 - 17 - 18 - 19 - 25 - 26 - AUGUST : 14 - 15 - 21pm - 22 - 23am. (am = morning, pm = afternoon). Combine the influence of a rising tide with these dates.
  The waxing and descending Moon can be suitable mainly when the soil is heavy.

MAIZE, SUNFLOWERS

- **Sow:** preferably when the Moon is ascending and if possible waning, with a rising tide.
- **Harvest:** when the Moon is ascending, avoiding dates favourable to 'leaf' plants.

OILSEED RAPE

- **Sow:** preferably when the Moon is ascending and if possible also waning.
- **Harvest:** when the Moon is waning and ascending, avoiding 'leaf' dates.

MILLET

- **Sow:** when the Moon is ascending and if possible waxing.
- **Harvest:** when the Moon is waxing and if possible ascending, but avoid 'leaf' dates.

Animal husbandry

COVERING: for cows and mares, the best time is between the first quarter and the Full Moon in Fire and Earth signs and, if possible, with the Moon in the descendent. For ewes, the period between the first quarter and the Full Moon during Earth and Air signs is best, with an ascendant Moon if possible. Avoid the perigee and the lunar nodes if possible.

TRIMMING THE FEET OF LIVESTOCK: carry out this task in the signs of Taurus, Leo, Virgo, Sagittarius or Capricorn when the weather is calm or wet and, if possible, when the Moon is waxing. Avoid stormy or icy periods.

SHOEING: choose the waning Moon, particularly when it is in Earth or Air signs.

WORMING: this should be done two or three days before the New Moon or, if this proves impossible, two or three days before the Full Moon.

CLEANING & DISINFECTING COWSHEDS: this will be most effective when the Moon is waxing and in a Fire or Earth sign.

MATING RABBITS: results will be best between the New Moon and the first quarter.

MALE OR FEMALE CHICKS: eggs collected around the time of the last quarter will produce mainly male chicks. For female chicks, eggs should be collected around the first quarter. For best results, incubate the eggs so that they hatch between the New Moon and the first quarter.

WHEN TO START OUTDOOR GRAZING: choose the following correlations of signs and constellations: a - b - c* - d* - k* - L* - m - n - s* - t* - x (see page 14). For example, **MARCH:** 1* - 2* - 3* before 17:00 - 11* - 12* - 20 before 10:00 - 21* - 24 - 26 - 27 after 08:00 - 28am - 28pm* - 29* - 30* - **APRIL** : 7pm* - 8* - 9* before 09:00 - 10 - 16* - 20 after 08:00 - 21am - 22pm - 23 - 24 - 25* - 26*- **MAY:** 5* - 6am* - 7 - 8 - 13* - 14* after 14:00 - 15* before 10:00 - 18 - 20 - 21 - 22* - 23* - 24am*. (am = morning, pm = afternoon).

MOVING CATTLE: move cattle between the day after the first quarter and the day before the last quarter, to reduce damage to grass.

BRINGING CATTLE IN: cowsheds will be dryer in winter if cattle are brought inside during autumn when the Moon is passing through the constellation of Leo.

TO AVOID: In addition to the above, avoid turning out, moving or bringing in animals on the following days: **JANUARY** : 3 - 19 - 30pm - **FEBRUARY** : 8am - 14am - 18 - 19am - 21 before 14:00 - 26pm - **MARCH** : 7 -13 - 14pm - 19 -20 after 15:00 - 27 before 08:00 - **APRIL** : 3pm - 9pm - 11 - 17 - 18 - **MAY** : 6pm - 14 before 14:00 - 15 between 10:00 and 14:00 - 17 - 19am - **JUNE** : 2 after 17:00 - 10pm - 15 - 24 before 09:00 - 30 before 07:00 - **JULY** : 7pm - 15 - 21pm - 27 - **AUGUST** : 74am - 9am - 13 - 23 after 14:00 - **SEPTEMBER** : 5am - 12 - 14pm - 19 after 17:00 - 28am - **OCTOBER** : 2 - 12 - 26 - **NOVEMBER** : 8 after 14:00 - 10 - 22 after 14:00 -**DECEMBER** : 6 -10 -23 - 26pm. (am = morning, pm = afternoon).

Beekeeping

BEES

For the best results, work with bees during the following constellations.

During Fire constellations to benefit the queens and increase the production of honey.

During Air constellations to benefit the queens and to encourage the development of young bees.

During Earth constellations to maximise the cell-building instinct. Honey collected during Earth constellations will set more quickly.

Avoid all work with bees during Water constellations – these periods have an adverse effect on the hives and honey.

Extract honey in a Fire or Air constellation.

As a general rule do most of your beekeeping when the Moon is in a Fire or Air constellation and avoid the perigee, lunar nodes and Water constellations.

Winegrowing

PRUNING GRAPE VINES

Prune grape vines when the Moon is both descending and, if possible, waxing, with a rising tide. Avoid the lunar nodes and when the Moon is squared with Saturn (☾ □ ♄).

DISBUDDING

Disbud when the Moon is waning and, if possible, descending, with a receding tide. Disbudding with the Moon sextile or trine Sun or Mars (☾ ✶ △ ⊙ ♂) should reduce the growth of suckers (see pages 12–14).

POLLARDING AND CUTTING BACK

Carry out this work when the Moon is descending, with a receding tide and, if possible, with one of the planetary aspects mentioned above.

HARVESTING GRAPES

An ascending Moon will give a better yield but the wine produced will keep better when the Moon is descending. To improve the quality of the wine and reinforce its character, harvest the grapes according to the constellations. Choose dates favourable to 'fruit' plants for fruity wines, 'root' days for vines grown where the soil plays an important part in the flavour, and 'flower' days for more floral wines. Avoid lunar nodes: they can hinder the development of aromas.

Cider making

DECANT

Decant cider when the Moon is waxing and descending if possible, and with a rising tide. Choose days when the weather is calm, preferably early in the morning with dew on the ground.

BOTTLE

Bottle cider when the Moon is waning and descending, preferably in the constellations of Virgo (♍) or Leo (♌) and, if possible, with a rising tide. Avoid days when the Moon is squared with Mars, Jupiter, Saturn or the Sun (☾ ◻ ♂ ♃ ♄ ☉) – see pages 12–14. Also avoid stormy, windy and very cold weather.

Beer making

WHEN TO START

Make your beer when the Moon is descending and, if possible, waxing, with a rising tide. Moon–Venus aspects (☾ ☍ ☌ △ ✶ ♀), except square, are very favourable. Avoid lunar nodes, the perigee and dates when the Moon is squared with Mars, Jupiter, Saturn or the Sun (☾ ◻ ♂ ♃ ♄ ☉) – see pages 12–14. For the best results use hops picked when the Moon is ascending and with the influence of a rising tide.

BOTTLE

Bottle beer when the Moon is descending and, if possible, between the last and first quarters. Choose a time influenced by a rising tide to help with the development of aromas. Choose an Air sign if possible, and avoid stormy, windy and very cold weather.

Forestry

If possible, only cut down trees after they have lost their leaves, but while the Sun is still in the descendent (before the winter solstice).

TIMBER

Cut timber when the moon is descending but avoiding the first and the last day of this phase. The wood will react differently according to the sign of the zodiac that the Moon is in.

The day on which wood is both cut and used is very important.

- To avoid **parasites** and **rotting**, cut wood in the following signs: Leo ▢, Virgo ▢, Scorpio ▢ or Sagittarius ■.
- To prevent wood from **warping and splintering**, cut it during Libra ▢ or, better still, in Scorpio ▢.
- Cut wood to be used for **frameworks for buildings and other structures** in Leo ▢, Virgo ▢, Libra ▢, Scorpio ▢ or Sagittarius ■.
- Wood for **tools** should be cut in Leo ▢, Virgo ▢ or Libra ▢.
- To prevent the softwood from coniferous trees from **flaking**, cut it in Scorpio ▢ or Sagittarius ■.
- **The superposition of the sign and constellation of Leo (♌) is a good time for cutting wood**, for example 02:00 to 23:20 on 18 November.
- When the sap is flowing, leave a felled tree for several weeks before removing the branches (the leaves draw out the sap).

Useful tip: to prevent wood from warping when you are using it (for frames, cladding, etc.), choose these dates ▭, especially those marked with stars (★).

In the table below, the signs are represented by colours for easier reference:

Gemini: ■ Cancer: ■ Leo: ▢ Virgo: ▢ Libra: ▢ Scorpio: ▢ Sagittarius: ■

Those planetary aspects that will improve the wood are shown by a star. Those that are harmful are coloured in red ■.

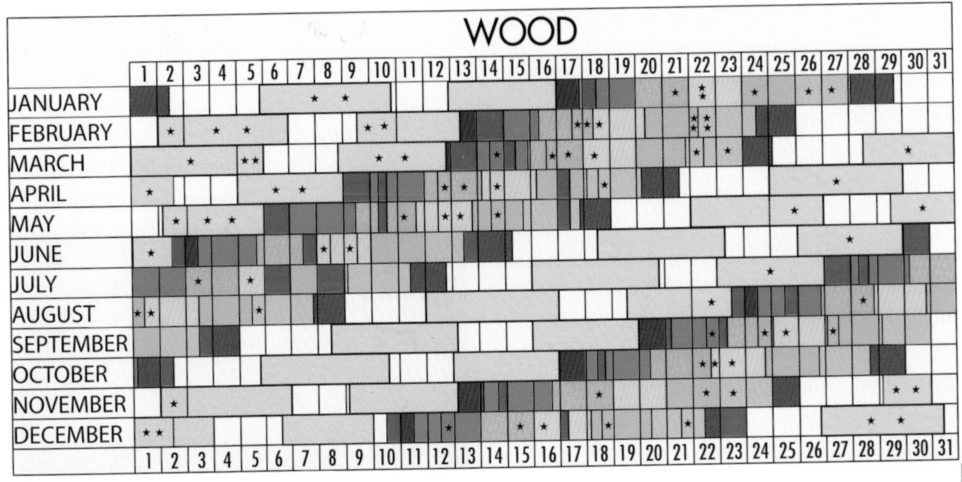

FIREWOOD: cut firewood when the Moon is descending, avoiding days close to the New Moon. Stack firewood when the Moon is descending, if possible between the first and last quarter – it will dry better then.

CHRISTMAS TREES: cut Christmas trees when the Moon is ascending and waxing to stop them shedding their needles too quickly. The following aspects will strengthen this effect – Moon sextile, trine or square Sun or Venus (☾ ✶ △ □ ☉ ♀) – see pages 12-14.

Miscellaneous

WORKING ON BOATS ON DRY LAND: carry out painting work when the Moon is in the descendant during periods (a, b, c, i, j, k). The following aspects can be beneficial: Moon sextile or trine Sun (☾ ✶ △ ☉) and Moon conjunction Jupiter (☾ ☌ ♃). Avoid the nodes and Moon square Sun (☾ □ ☉). Return boats to the water when the Moon is waning, if possible between the last quarter and the New Moon and when the Moon is in an Earth or Air sign.

CHIMNEYS: to improve the 'draw' of a chimney in autumn, light the fire during an ascending Moon, on a day favourable to 'fruit' and, if possible, between the Full Moon and the final quarter. Sweep chimneys when the Moon is in the constellation of Pisces between the New Moon and the first quarter.

MOSS ON TILES, SALT DEPOSITS ON BRICKWORK, MOULD: remove when the Moon is waxing and ascending, avoiding Water signs and constellations. The nodes, the perigee and the Moon squared with the Sun and Saturn (☾ □ ☉ ♄) are all favourable, and a drying wind will help.

DRAIN SOIL: when the Moon is ascending.

PATHS AND COURTYARDS: to work on paths and courtyards, choose (the best are marked *): **JANUARY:** 18am - **FEBRUARY:** 14* - **MARCH:** 13 - 17*- **APRIL:** 9 - 13pm* **MAY:** 6pm - 11*before 15:00 - **JUNE:** 7* - **JULY:** 4*- 31*after 14:00 - **AUGUST:** 1 before 10:00 - 28*.
(am = morning, pm – afternoon) (in the UK).

TRENCHES: the sides of trenches dug or cleaned out when the Moon is ascending may erode more quickly, especially if the soil is light, but should not fall in completely. Trenches dug when the Moon is descending tend to erode and collapse. Choose a time close to the Full Moon to help trenches last longer.

SPRINGS: begin work to collect spring water on the day of the New Moon and when the Moon is ascending (water is lost when it is descending).

MUSHROOMS: usually emerge a few days after the New Moon and are more prolific when the Moon is waxin

SILVER BIRCH SAP: collect silver birch sap when the Moon is ascending, avoiding the nodes and period (f).

PRESERVE FOOD: when the Moon is waning.

SAUERKRAUT: pick cabbages and make sauer-kraut when the Moon is ascending, avoiding the same periods as for bread.

JAMS: jams with a low sugar content will keep better if made when the Moon is waning.

SOURDOUGH BREAD: this will rise well and be extra delicious when the Moon is ascending. If the Moon is descending when baking, ensure that it is also waxing. Avoid nodes, the perigee, the periods (f, o, w), Water signs and constellations, if possible, and the Moon squared with the Sun and Saturn (☾ □ ☉ ♄) – see pages 12-14.

CLEANING AND MAINTENANCE WORK: will be more effective when the Moon is waxing.

CLOTHES: put clothes away for storage when the Moon is in conjunction with Venus (☾ ☌ ♀) – see pages 12-14.

WOODEN POSTS: to make them last longer, cut them when the Moon is new (or just before) and insert them in the ground when the Moon is in the constellation of Leo. If they are made of softwood, you can also burn the point a little to make it harder.

LAWNS: sow lawns when the Moon is waning and descending on a day favourable to 'leaf' plants or, if necessary, a 'root' date. Aerate lawns when the Moon is ascending on a 'flower' date. To help slow down growth, cut lawns when the Moon is in the constellation of Gemini and Libra or, if necessary, in Virgo or Aquarius. To promote growth, cut lawns in Pisces or, if necessary, in Cancer or Scorpio.

LIVING WITH THE MOON

Do not use the information contained in this chapter as a substitute for medical advice. If you think you have a medical problem, consult your doctor.

Influence of the phases

NEW MOON

• The renewal of physical energy is at a minimum during this phase.
• This is a good time to start eliminating toxins or to tackle a bad habit.

WAXING MOON

• We express ourselves more easily and freely.
• The body makes the most effective use of the nutri - ents that it receives during this phase.
• If you tend to put on weight, this is a time to be particularly frugal..
• This is an ideal time to learn new skills or to start new projects.

FULL MOON

• Our physical energy levels are restored during this phase.

• When the nights are lighter sleep can be interrupted, and insomnia can be an issue.
• The extra energy and excitement caused by the Full Moon can be channelled into physical activities.
• Now is a good time to undertake a project.
• Some of the effects of the Full Moon and the New Moon can be felt for two or three days.

WANING MOON

• This is a good time for introspection, reflection and spirituality.
• Choose this phase to follow a detox programme; it promotes the elimination of toxins from the body.

Harvesting plants

Waxing Moon: plants harvested during the Full Moon period will have more vitality, but their medicinal properties are weaker than when the Moon is waning.

Waning Moon: plants tend to become dry but their fragrance heightens as the New Moon approaches. Plants picked during this phase will retain more of their medicinal properties.

FOR MAKING INFUSIONS

In the morning while the Moon is ascending: before midday, the Sun's influence is felt mainly on the aerial parts of a plant. It is the ideal time for harvesting leaves, flowers, fruits and seeds, and the effects are enhanced when the Moon is ascending.

In the evening while the Moon is descending: after 15:00 the Sun's influence stimulates activity in the lower parts of plants. This is the time to harvest roots. The effect is enhanced when the Moon is descending.

The following days are harmful: avoid harvesting plants when the Moon is passing through the lunar nodes when ascending or descending or when it is at its perigee (see red zones).

Colt's foot

MEDICINAL PLANTS

Harvesting flower heads (borage, St John's Wort, meadowsweet, etc.): when the Moon is waning and, if possible, also ascending, in an Air or Earth sign. Early morning is best. The following aspects are favourable: Moon conjunction, sextile or trine Sun, Venus and Mercury (☾ ☌ ✶ △ ☉ ♀ ☿) – see pages 12-14.

Borage

Harvesting roots (gentian, marsh mallow, dandelions, etc.): when the Moon is descending and, if possible, also waxing, in an Air or Earth sign. Late afternoon is best. The following aspects are favourable: Moon conjunction, sextile or trine Sun, Venus and Mercury (☾ ☌ ✶ △ ☉ ♀ ☿) – see pages 12-14.

Harvesting from trees (hawthorn, pine nuts, elderberries, etc.): when the Moon is ascending and, if possible, waxing, with the influence of a rising tide and in an Air or Earth sign. The following aspects are favourable: Moon conjunction, sextile or trine Sun, Venus and Mercury (☾ ☌ ✶ △ ☉ ♀ ☿) – see pages 12-14.

Lime blossom

Hairdressing

The table below summarizes for quick reference the best dates and times of day on which to cut hair.

- a good day for cutting hair to **slow down hair loss**
- a good day for cutting hair to make it **thicker and stronger**
- (★) **the number of stars** indicates the strength of influence
- **not advisable** in any circumstances. Also, to a lesser extent, it is not recommended on days marked with the sign (–)
- a vertical line indicates the importance of a particular part of the day. The example shows a morning favourable for reducing hair loss and a neutral afternoon
- days divided horizontally are favourable for the influences represented by the two colours

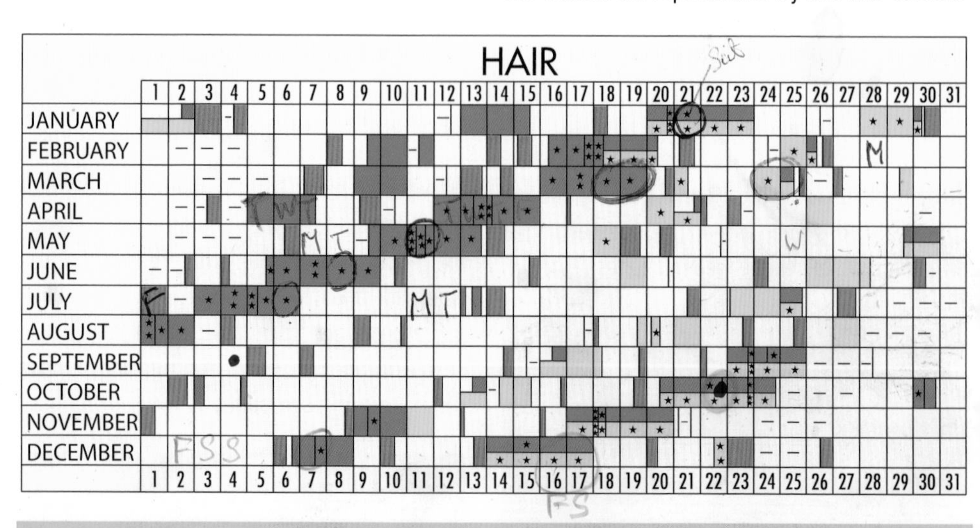

The dates marked in this table (■, □), especially the starred ones, are highly favourable for hair. Take note of them, even if you have no particular hair problems.

- For more **manageable hair**, choose dates marked in green when the Moon is waxing.
- For a **perm**, choose the superpositions of sign and constellation ♉ ♉ (j) or ♍ ♍ (s) (see calendar pages 56-79), if possible with a waning Moon. It sets better when the Moon is waxing but the hair suffers more damage.
- Hair **coloured** when the Moon is ascending and waxing will achieve better results.
- To help hair **grow quickly**, cut it when the Moon is ascending and waxing.
- If you tend to have **split ends** and/or you want your hair to grow less quickly, we recommend that you have it cut on the following days:

am = morning pm = afternoon

JANUARY : 1 - 2 before 10:00 - 20 before 15:00 - 20* after 15:00 - 21 - 22 - 23 - 24 - 25am* - 25pm - 26am - 27am* - 27pm - 28pm - 29 before 16:00 - **FEBRUARY** : 18 after 09:00 - 19* - 20 - 22*- 23 - 25 - **MARCH** : 20am - 21am* - 21 until 15:00 - 22 - 23 - 24 - **APRIL** : 18 - 19 - 20pm - 21 before 14:00 - **MAY** : 18 - **JULY** : 28 - **AUGUST** : 24am - 24pm*- 25am - 26 - 27 - **SEPTEMBER** : 20 - 21 - 22 after 09:00 - 23 - 24 - 25 - **OCTOBER** : 17 - 18 - 19am - 20 - 21 - 22 - 23 - 24 - **NOVEMBER** : 13pm - 14 - 15 - 16 after 09:00 - 17 - 18am* - 18pm - 19 - 20 - 22 - 23 - **DECEMBER** : 11 - 12 - 13am - 14 - 15 - 16 - 17 - 18 - 19 - 20pm - 21 - 22pm.

Depilation

The removal of unwanted hair is best performed when the Moon is descending; if combined with the waning Moon, regrowth will be slower. Slow down regrowth even more by choosing certain planetary aspects.

A summary indicating the best dates is shown in the table below. The dates are divided into four categories:

fairly good days | good days | very good days | ★ excellent days

a vertical line indicates a particular part of the day

DEPILATION

	1	2	3	4	5	6	7	8	9	10	11	12	13	14	15	16	17	18	19	20	21	22	23	24	25	26	27	28	29	30	31
JANUARY																															
FEBRUARY																					★										
MARCH																				★			★								
APRIL																															
MAY																															
JUNE																														★	
JULY							★																				★				
AUGUST																							★		★						
SEPTEMBER																					★	★									
OCTOBER																		★	★							★					
NOVEMBER																★				★	★										
DECEMBER													★							★			★								
	1	2	3	4	5	6	7	8	9	10	11	12	13	14	15	16	17	18	19	20	21	22	23	24	25	26	27	28	29	30	31

Skincare

SKIN CLEANSING

(Blackheads, impurities, etc.) Treatment works best if carried out when the Moon is in the descendent and in an aspect with Venus (☌, ✶, □, △) – see pages 12-14.

NOURISHING THE SKIN

Moisten the skin with dew before applying any cream for best results. Walking barefoot in the dew is good for your general health.

- **Face packs,** moisturizing and revitalizing treatments: the right time depends on skin type.
- **Dry and combination skins:** treatments will be more effective when applied during a waxing Moon. Avoid the following signs: Aries, Cancer and Capricorn.
- **Greasy skin:** when the Moon is waning and trine and, especially, sextile with the Sun (☾ △ ✶ ☉) – see pages 12-14. The skin will absorb the products a little less well but the cleansing action will be more effective. Perform during a waning Moon to help prevent dilation of the pores, making the skin firmer and more attractive.

Marigold

Warts

To remove warts with the sap from plants such as greater celandine, marigold and the fig tree...

Start when the Moon is descending, preferably during the last quarter, and when it is in the superposition of the following signs and constellations:

- Leo/Leo (q),
- Gemini/Taurus (k)

(see pages 56-79)

Celandine

Nails

For stronger nails, cut or file them when the Moon is waning and in an Earth or Air sign. Avoid the nodes.

Reduce the risk of in-growing nails by cutting them when the Moon is in the following signs: Leo, Virgo, Sagittarius, Capricorn and Aquarius, avoiding the lunar nodes.

Corns and callouses

Remove these when the Moon is waning and in the sign of Capricorn or Taurus. Another suitable time is when the Moon is waxing and in the sign of Virgo or Libra, though the effects are likely to be less successful.

Fasting

Fasting can be of great benefit to most people, but it is advisable to get permission from your doctor before embarking upon your first fast. Monitoring is essential for long fasts.

The New Moon is the best time to carry out a one-day fast, as your body will be cleansed more effectively. The New Moon is also a good time to begin fasts of up to a week. If you wish to fast for longer, begin when the Moon is descending and sextile with the Sun (☾ ✶ ☉) – see pages 12-14.

Detoxing

Start taking plants that help to detox your system through the liver when the Moon is in the sign of Virgo, and plants that work through the kidneys when it is in the sign of Libra. If possible, choose a time when the Moon is waning and sextile or trine with the Sun (☾ ✶ △ ☉) – see pages 12-14.

Treating worms

Carry out treatment to eradicate worms two or three days before the New Moon or, failing that, two or three days before the Full Moon.

Eating

Digestive problems ?

Key

Use your intuition to choose foods that most benefit you, and always eat in moderation.

A few pointers

- Meals should be eaten in a calm, relaxed atmosphere. Chew well and savour them to the full.
- Eat fruit and vegetables in season and grown locally (their energy is best adapted to our bodies).
- Give priority to vegetables (steamed or stir-fried in a wok) and fruit.
- Remember to eat sprouting seeds, seaweed, fibres, raw fruit (in between meals), and non-processed foods; use first cold pressed oils.
- Avoid combining starchy and protein foods in the same meal. Digesting them simultaneously makes conflicting demands on your stomach.
- Avoid in-between meals snacking, eat in moderation preserved meats, smoked products, soya beans and oats.
- If in doubt about a specific food, stop eating it for at least three weeks in order to monitor its effect on your body when you resume eating it.

Frequent issues

Excess of mucus: in order to help the liver when tired or stressed, limit or even avoid completely all dairy products.

Stomach acidity: meat and poorly-processed grains disrupt body equilibrium through acidification. Stress and lack of oxygen act similarly. Give priority to vegetables and in order to avoid deficiences, eat other foods in moderation.

Meals and the Moon

- **When the Moon is waxing, nutrients are absorbed more easily.**
- **When the Moon is waxing and ascending, digestion requires less energy.**
- **People who suffer from constipation should avoid foods that contain little water (such as cereals) during Fire signs and the lunar nodes.**

In order to conserve your vital energy, avoid the foods in the table below during the times indicated with letters (see p.10 and the calendar on pages 56-79). During those times, they take more energy to be digested properly, in particular for people with slow digestion.

Beef in sauce: c-f-g-o-u-w - Mutton in sauce: d-e-f-j-r-s-u - Pork in sauce: c-d-h-o-q-s-t - Poultry in sauce: b-i-j-q-r-s-u-v-w - Preserved meats: c-d-h-i-j-L-q-r-s-t-w-x-y - Oily fish: b-c-e-f-j-k-L-o-p-y Smoked fish: a-b-c-d-h-i-j-k-L-o-p-q-u-v-y-z - Oysters and raw shellfish: h-m-n-r-u-v-y - Eggs: f-g-i-L-m-n-q-r-u-y - Dried beans: e-j-m-n-q-r-y - Lentils: b-c-e-m-n-q-s-y -Raw celeriac: i-j-k-l-q-w-y - Raw cabbages: c-d-i-j-m-n-v-w-x - Raw cauliflowers: e-f-g-h-m-n-o-q - Cucumbers, melons: b-c-d-f-s-t-u-v-w-y-z - Radishes: a-i-j-k-q-r-s-x - Salads (curly endive, escarole): e-i-j-k-L-m-n-r-s-y-z Raw tomatoes: b-i-m-n-o-q-r-y

How to use the calendar when living outside the United Kingdom

Times throughout the book are given in **Greenwich Mean Time** (30 October to 26 March) or **British Summer Time** (27 March to 29 October), as appropriate. **For those in the United Kingdom and Ireland, all times given, both summer and winter, are those shown on your watches.** The number of hours to take off or add on (if any) for other countries appear on the world map on the next two pages. Use these not only to calculate the right times but also the right periods of the day.

Let's take the example of the compost table from page 29, repeated below: the first colour bar appears on the afternoon of 5 January - this means afternoon in London, when it is evening (same day) in east Asia and morning (same day) in west America. So when you use tables like these, remember to take into consideration where you live to determine when to take benefits from the moon cycles.

COMPOST

	1	2	3	4	5	6	7	8	9	10	11	12	13	14	15	16	17	18	19	20	21	22	23	24	25	26	27	28	29	30	31
JANUARY					★	★		★★★	★						★	★										★★					
FEBRUARY			★	★ / ★★	★						★	★										★ / ★★									
MARCH				★★ / ★	★						★	★		★								★★ / ★									★★
APRIL							★	★		★ / ★								★ / ★★									★ / ★★				
MAY				★	★		★ / ★									★★★ / ★									★						
JUNE		★	★									★★ / ★★									★★										
JULY									★ / ★★ / ★★									★★										★			
AUGUST					★ / ★							★	★																		
SEPTEMBER		★ / ★						★	★	★	★★★ / ★★	★																	★★ / ★★		
OCTOBER					★	★	★	★★★	★									★													
NOVEMBER		★	★	★★ / ★★★	★																		★★						★	★	
DECEMBER	★	★★ / ★						★	★											★							★	★	★★★	★	
	1	2	3	4	5	6	7	8	9	10	11	12	13	14	15	16	17	18	19	20	21	22	23	24	25	26	27	28	29	30	31

(★) the more stars, the more favourable the day

- a good time to add to or turn compost heaps that generate a high internal temperature (the majority of compost heaps)
- choose these dates for working on slower acting compost heaps that generate less heat (often preferred by farmers and gardeners who practise biodynamics)
- recommended for surface composting with non-rotted manure (broken down by worms);
- a vertical line dividing a date box indicates morning (left) or afternoon (right)
- a date box divided by a horizontal line indicates it is favourable for a certain type of composting (green for high temperature compost and orange for surface composting)

MANURE: spread manure when the Moon is descending, if possible on days marked orange (see the table above).

For tides, see page 80.

Calculating the time difference between where you live and London

Times throughout the book are given in **Greenwich Mean Time** (30 October to 26 March) or **British Summer Time** (27 March to 29 October), as appropriate. **For those in the United Kingdom and Ireland, all times given, both summer and winter, are those shown on your watches.** The number of hours to take off or add on for other countries appear in the map below – with some exceptions:

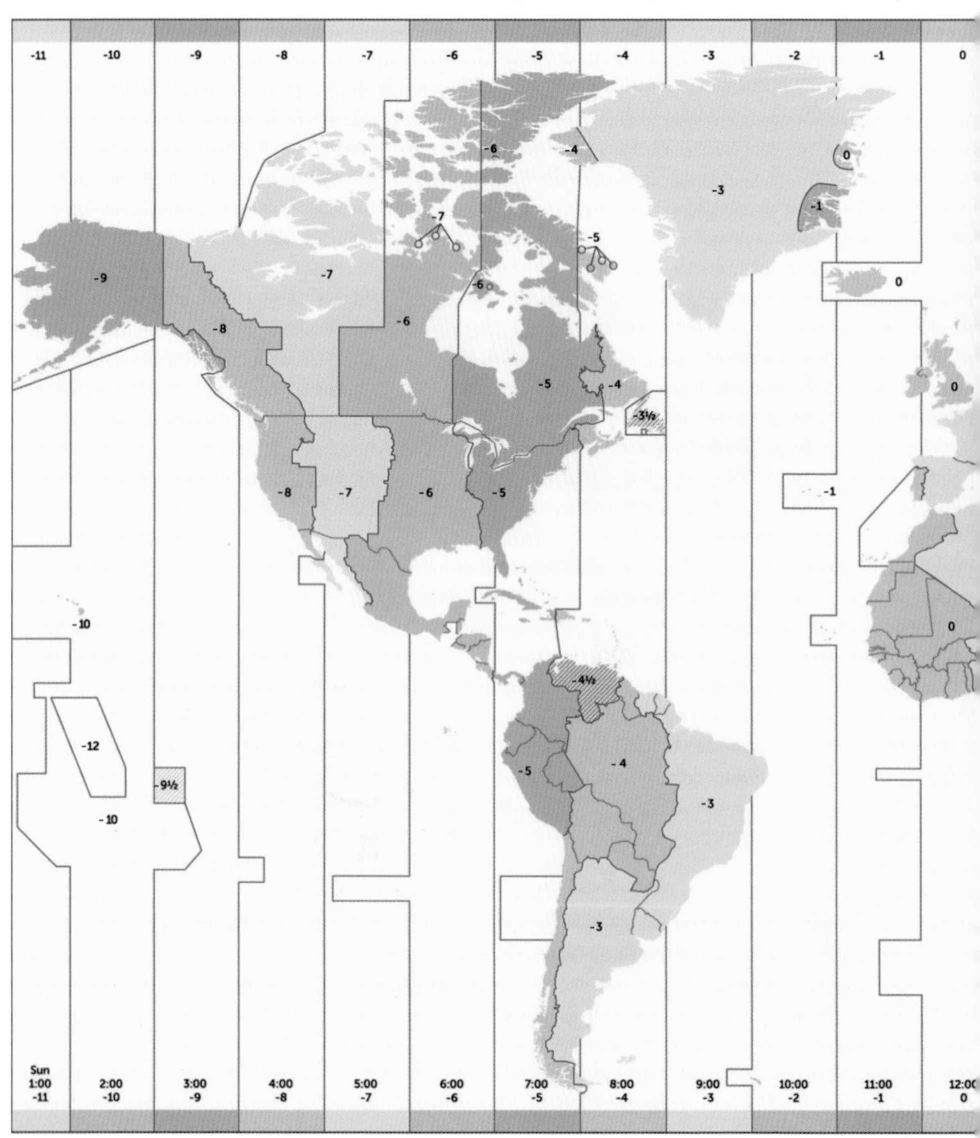

• USA and Canada: the switch to Daylight Saving Time (DST) does not happen on the same dates as in Europe — this means that between 13 and 26 March, as well as between 30 October and 6 November, you will need to reduce the time difference indicated on the map below by 1 hour.

• some countries do not observe Daylight Saving Time (DST), for instance Japan, Iceland, South Africa, Morocco, etc.: for these countries, use, the time difference as per the map below until 26 March and from October 30, and use the same figure minus 1 hour between 27 March and 29 October.

• New Zealand: add 13 hours before 26 March and after 30 October, add 12 hours between 27 March and 3 April and also between 25 September and 29 October, and add 11 hours the rest of the time (from 4 April to 24 September)

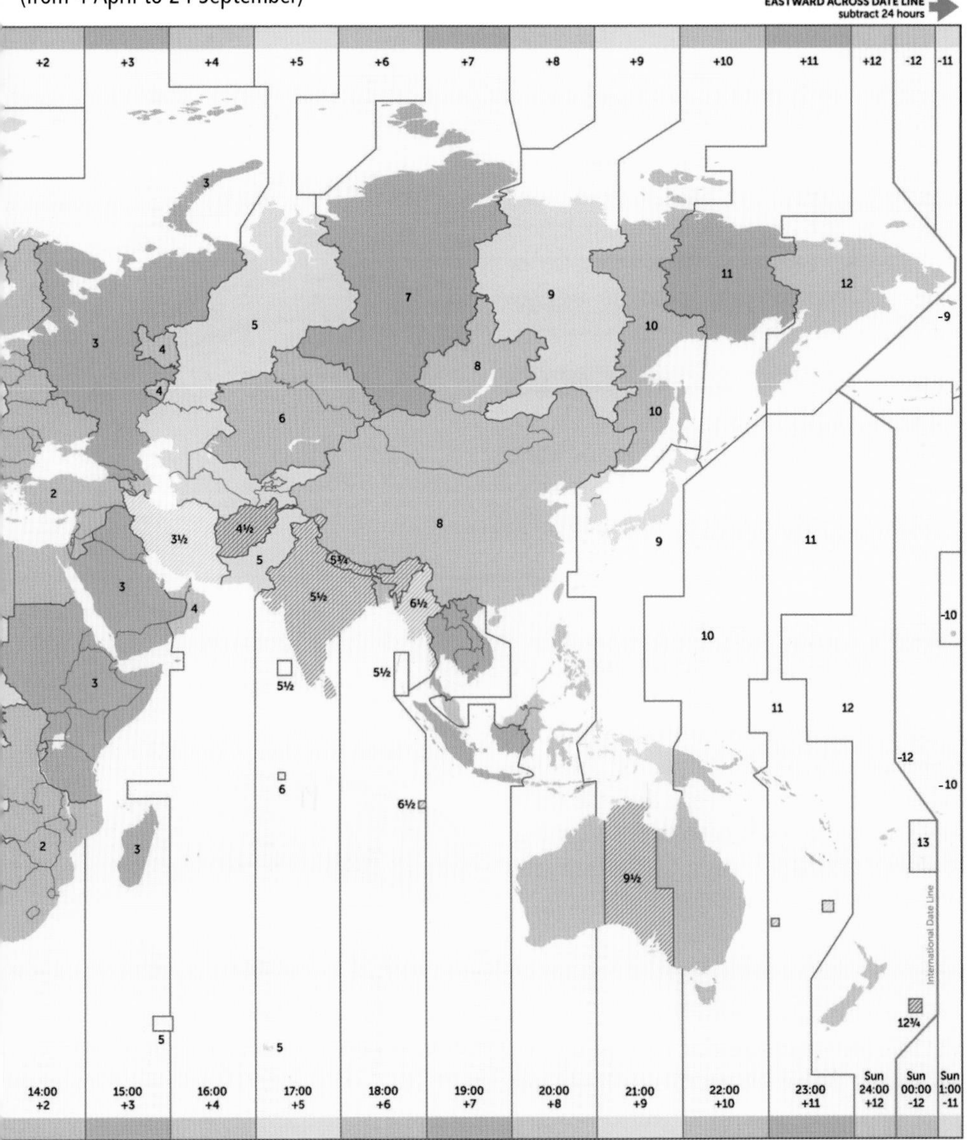

Understanding the calendar

On the calendar on pages 56-79, the phases of the Moon (New Moon ●, first quarter ◐, last quarter ◑, Full Moon ○) as well as its perigee P and its apogee A, are represented in the top band.

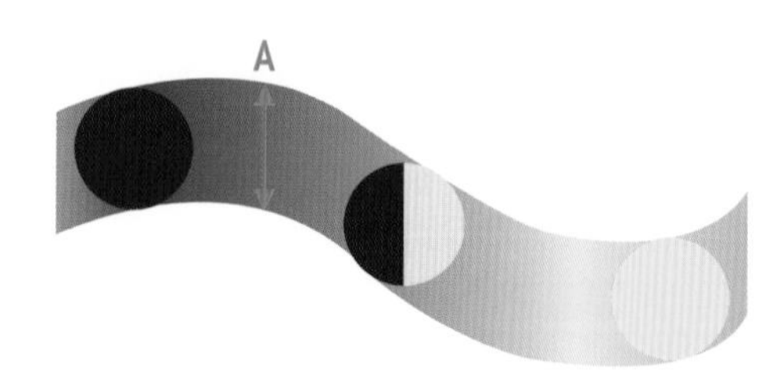

The band beneath shows the Chinese seasons and inter-seasonal periods. Beneath these are the bands indicating the signs and constellations of the zodiac, each of which is illustrated by a coloured strip that corresponds to the four elements:

FIRE | EARTH | AIR | WATER

The letters between these bands indicate where signs and constellations can be used in conjunction (see page 10).

The curve of the coloured bands represents the Moon ascending and descending and the black triangles indicate when the change occurs. When the bands curve upwards, the Moon is ascending and vice versa. This movement is also reflected by the thin coloured bands beneath the wider blue band with the dates – green for the ascending Moon and yellow for the descending Moon.

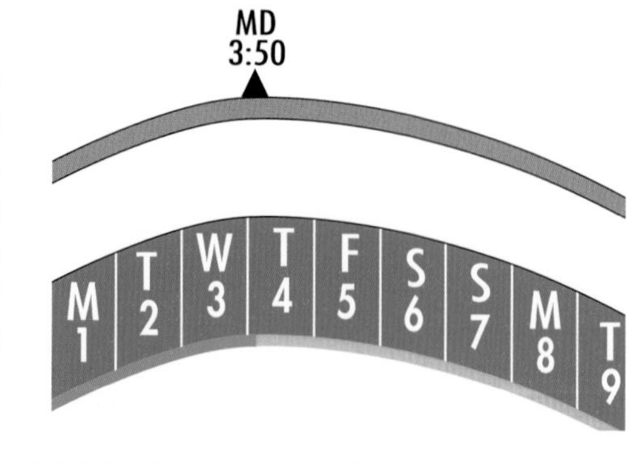

Cultivating the soil (particularly at depth), planting (for example, trees), planting out (especially plants with bare roots) and spreading compost or manure should be carried out when the Moon is descending, which is the best time for plants to take root.

The signs, letters and numbers in the white band are relevant to harvesting. The length of time is indicated by the vertical lines.

Ascending Moon:
harvest aerial parts of plant (green stars)
Descending Moon:
harvest parts of plant below ground (yellow stars)

The period for sowing different plants is indicated by the green band, which is in parallel with the constellations. The time when work should be carried out depends on why the plant is being grown (for its fruits, roots, flowers or leaves), avoiding the red zones. For example, choose a date favourable to 'leaf' plants for houseplants that do not flower and a 'flower' date for plants that do. Another example is to sow endives on 'root' dates and force their growth on 'leaf' dates.

The white band completes the information given in the green band. A summary of all the lunar and planetary effects on crops is given by indicating various zones ranging from 'harmful' (red band) to 'very favourable' (***) and also 'to be avoided' (—). Use these symbols in conjunction with the green band above for a clear and comprehensive understanding.

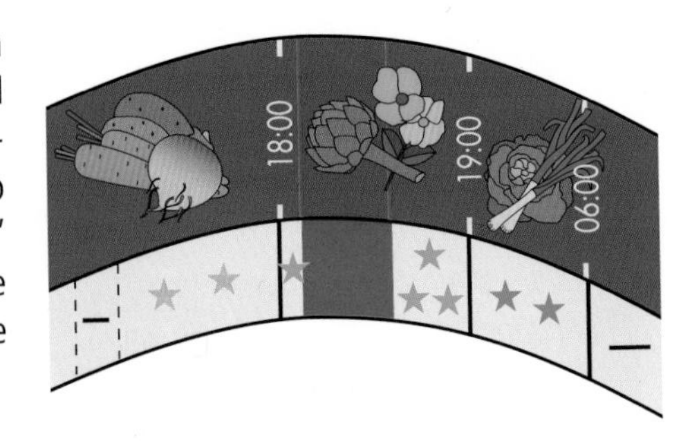

All the times mentioned are Greenwich Mean Time or British Summer Time, as appropriate. For other countries, see page 80.
Cut out the detachable page giving instructions on how best to use the calendar and use it as a bookmark to remind you to use the calendar every month.

JANUARY

3 January: the Earth is at its closest to the Sun (the perihelion) – about 147 million km (92 million miles).

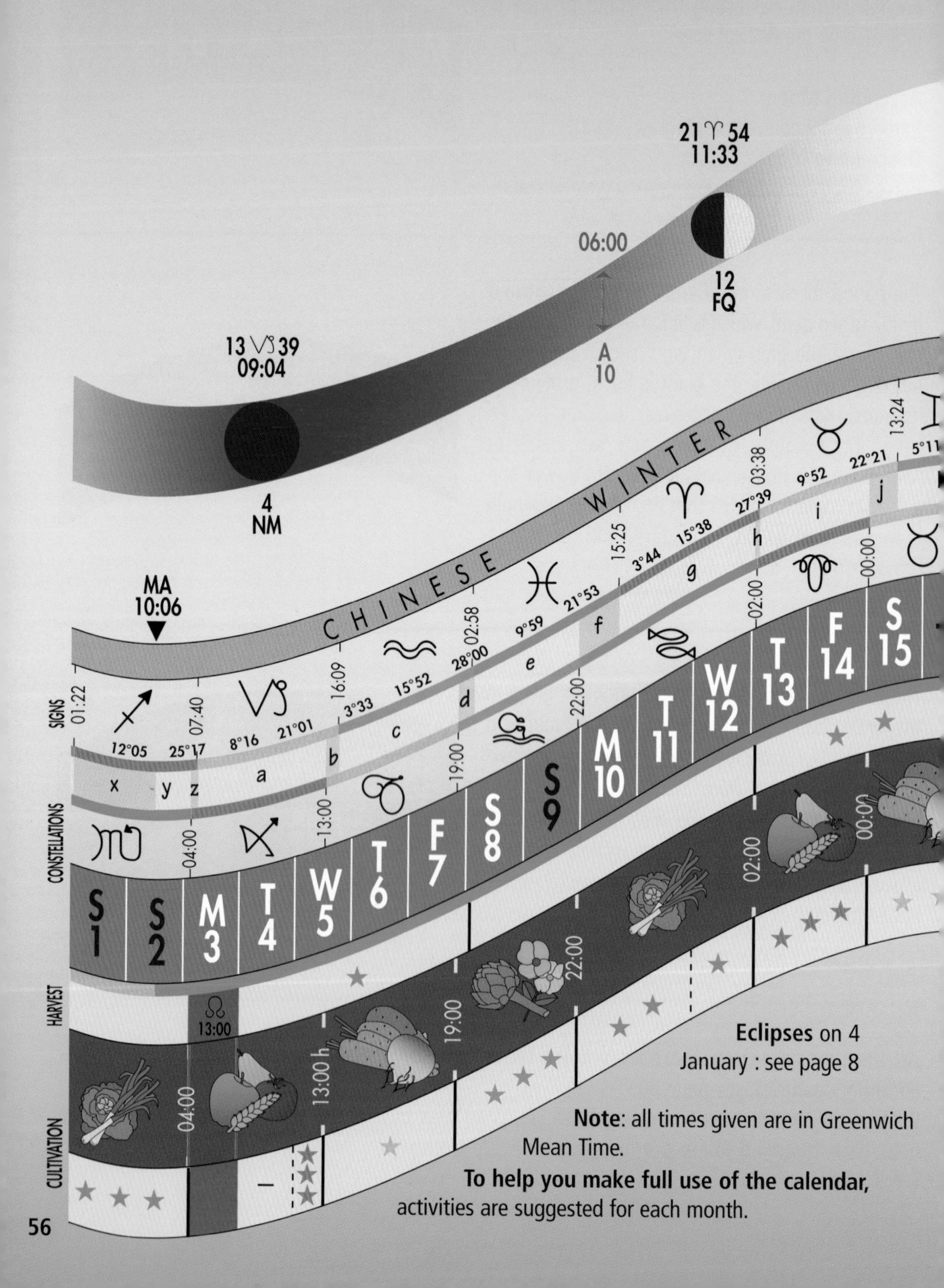

Eclipses on 4 January : see page 8

Note: all times given are in Greenwich Mean Time.

To help you make full use of the calendar, activities are suggested for each month.

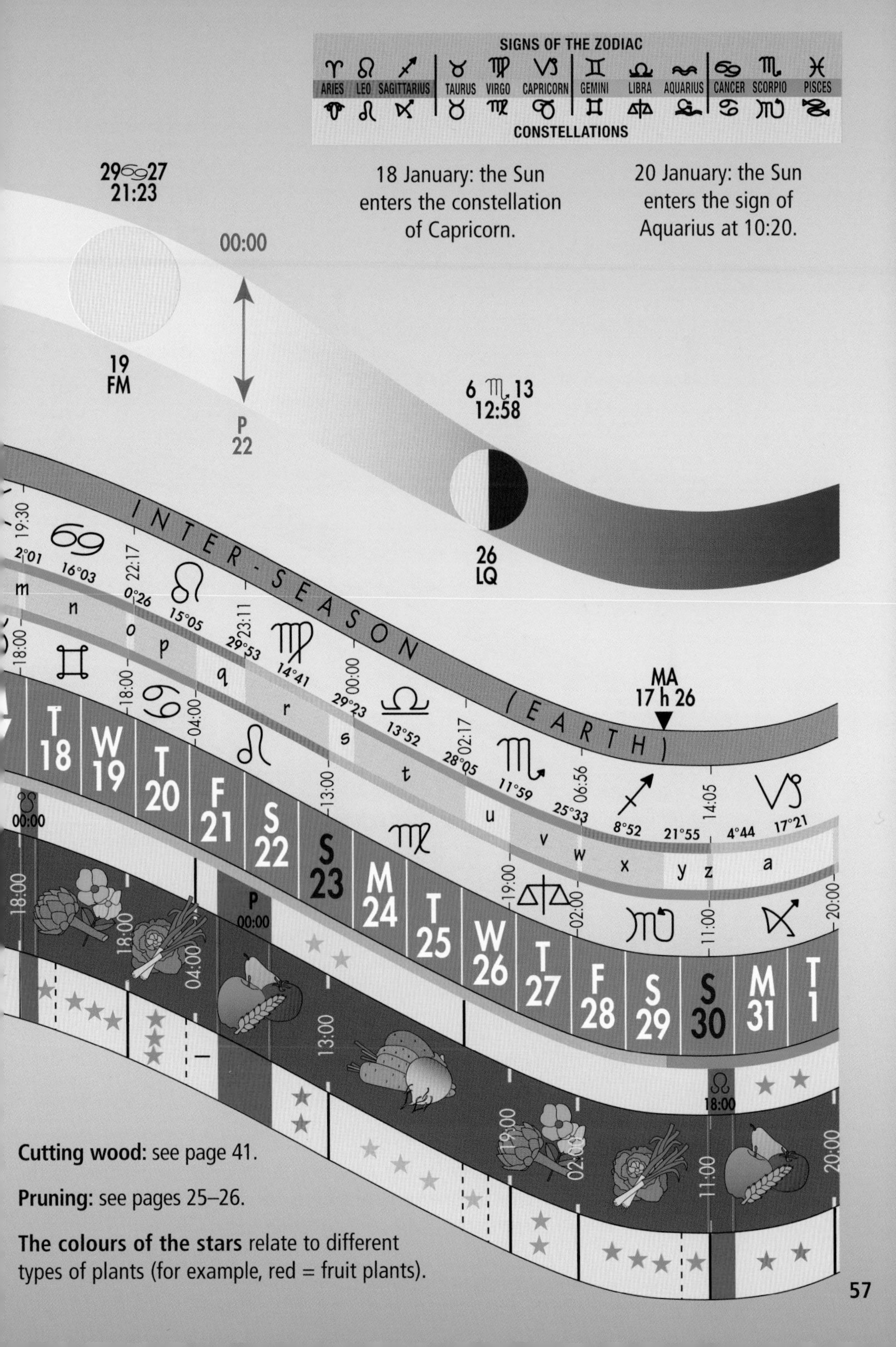

Cutting wood: see page 41.

Pruning: see pages 25–26.

The colours of the stars relate to different types of plants (for example, red = fruit plants).

FEBRUARY

3 February:
Chinese New Year.
year of the rabbit

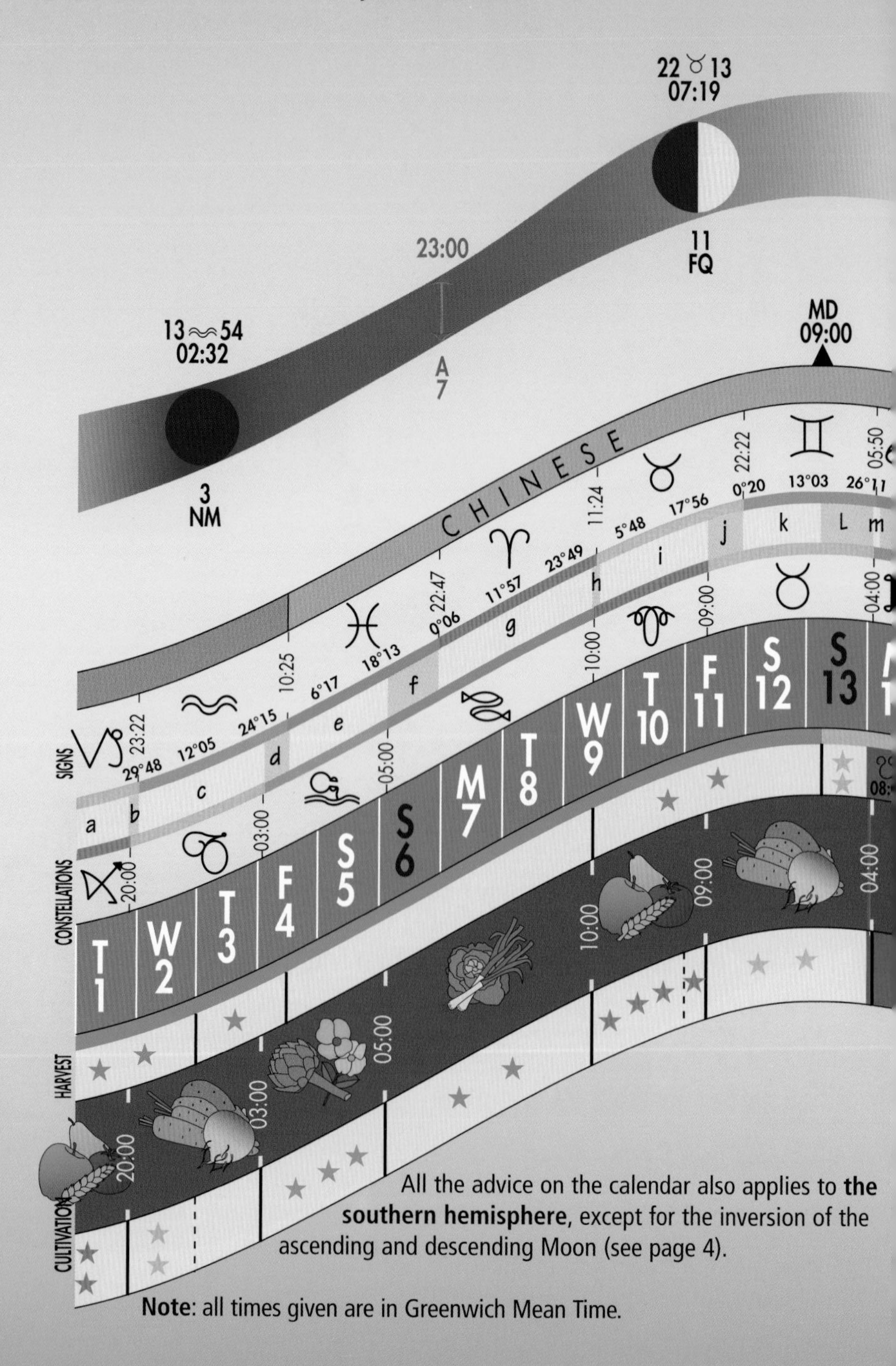

All the advice on the calendar also applies to **the southern hemisphere**, except for the inversion of the ascending and descending Moon (see page 4).

Note: all times given are in Greenwich Mean Time.

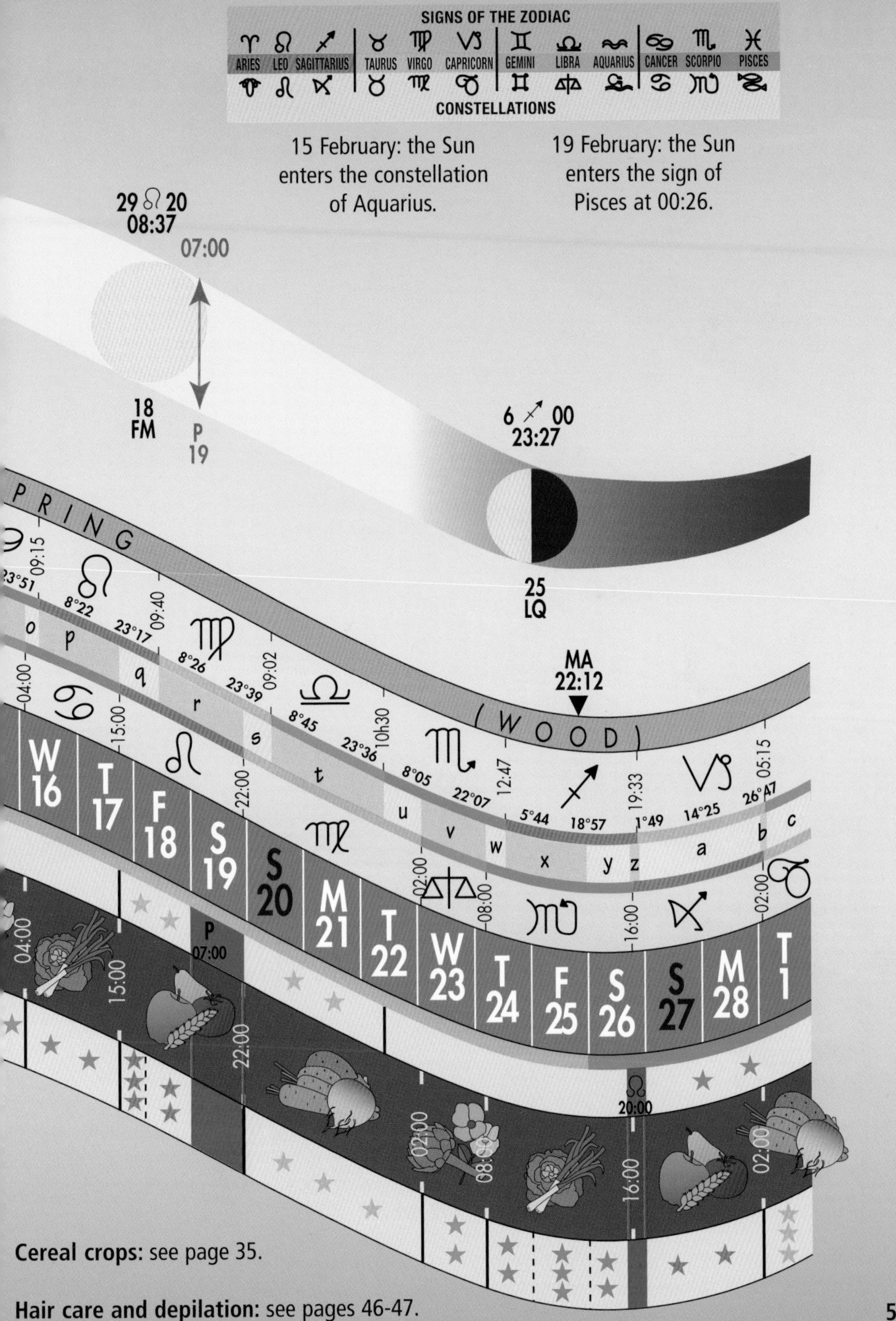

Cereal crops: see page 35.

Hair care and depilation: see pages 46-47.

MARCH

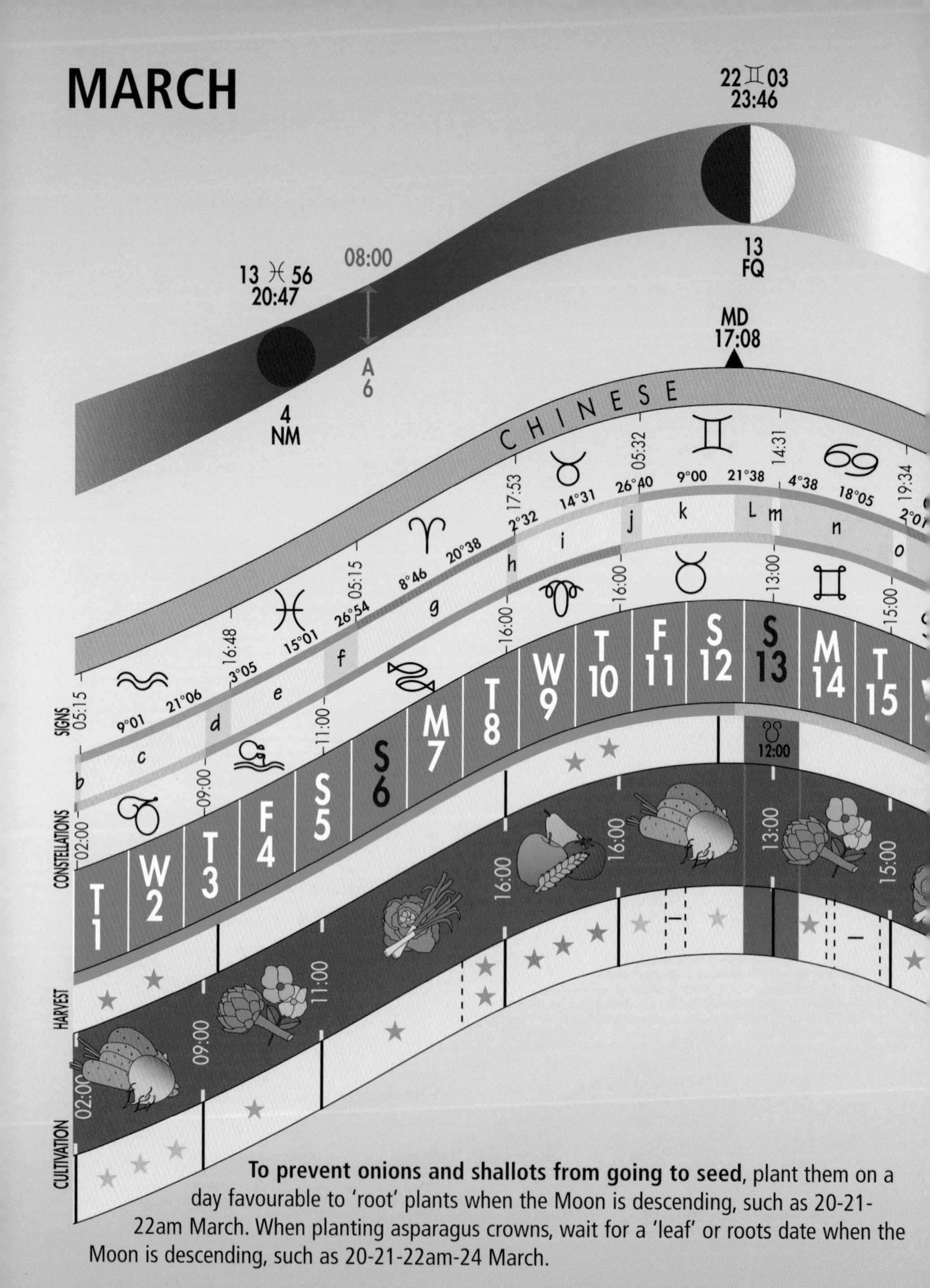

To prevent onions and shallots from going to seed, plant them on a day favourable to 'root' plants when the Moon is descending, such as 20-21-22am March. When planting asparagus crowns, wait for a 'leaf' or roots date when the Moon is descending, such as 20-21-22am-24 March.

March is a good month for **pruning** apple trees, but do not leave it any later. **Prune** when the Moon is descending, for example from 13 to 24 March. See page 25.

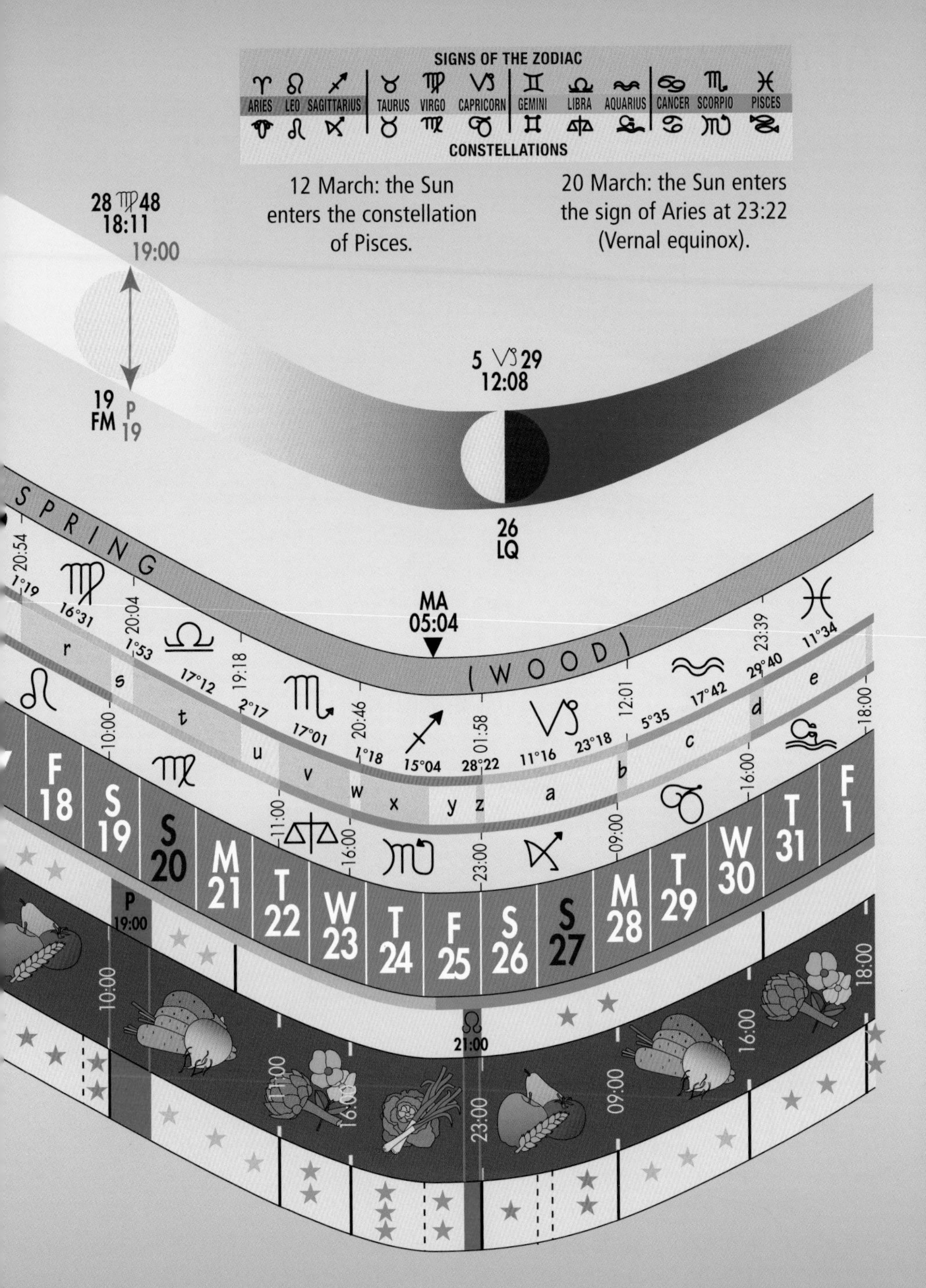

Note: all times given are in Greenwich Mean Time until 26 March. From 27 March, they are given in British Summer Time, thus taking into account the daylight saving change in the UK and Ireland.

APRIL

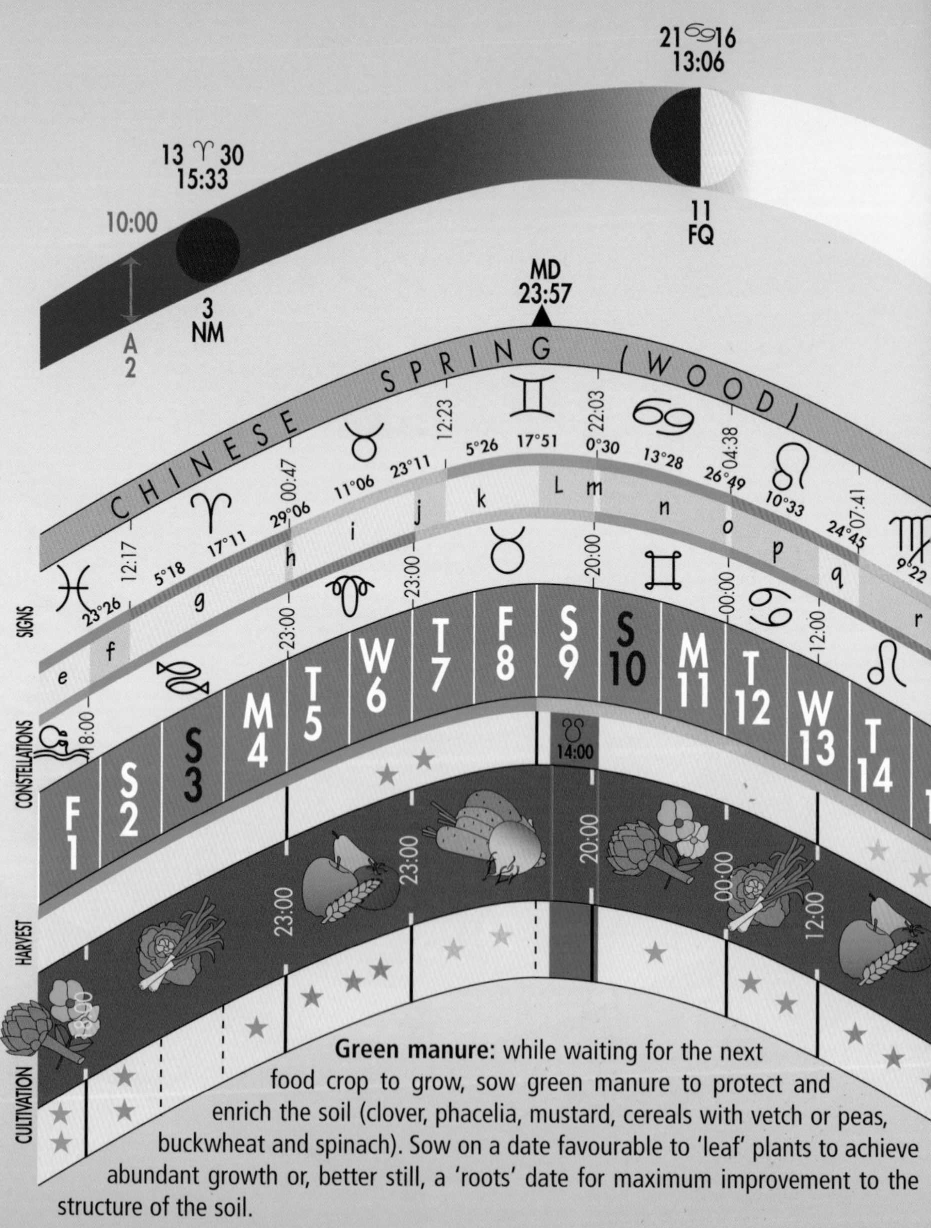

Green manure: while waiting for the next food crop to grow, sow green manure to protect and enrich the soil (clover, phacelia, mustard, cereals with vetch or peas, buckwheat and spinach). Sow on a date favourable to 'leaf' plants to achieve abundant growth or, better still, a 'roots' date for maximum improvement to the structure of the soil.

For plants that easily go to seed (lettuces, onions and so on), look for the appropriate constellation with the Moon in the descendent. For instance, 12-13 April for lettuces.

Grafting: see page 26.

Weeds: see page 33

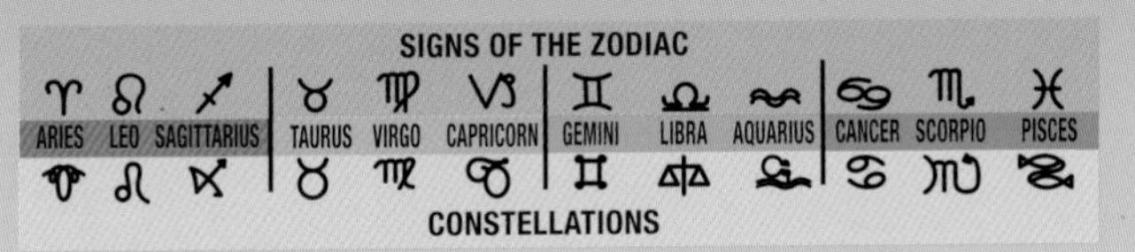

19 April: the Sun enters the constellation of Aries.

20 April: the Sun enters the sign of Taurus at 11:19.

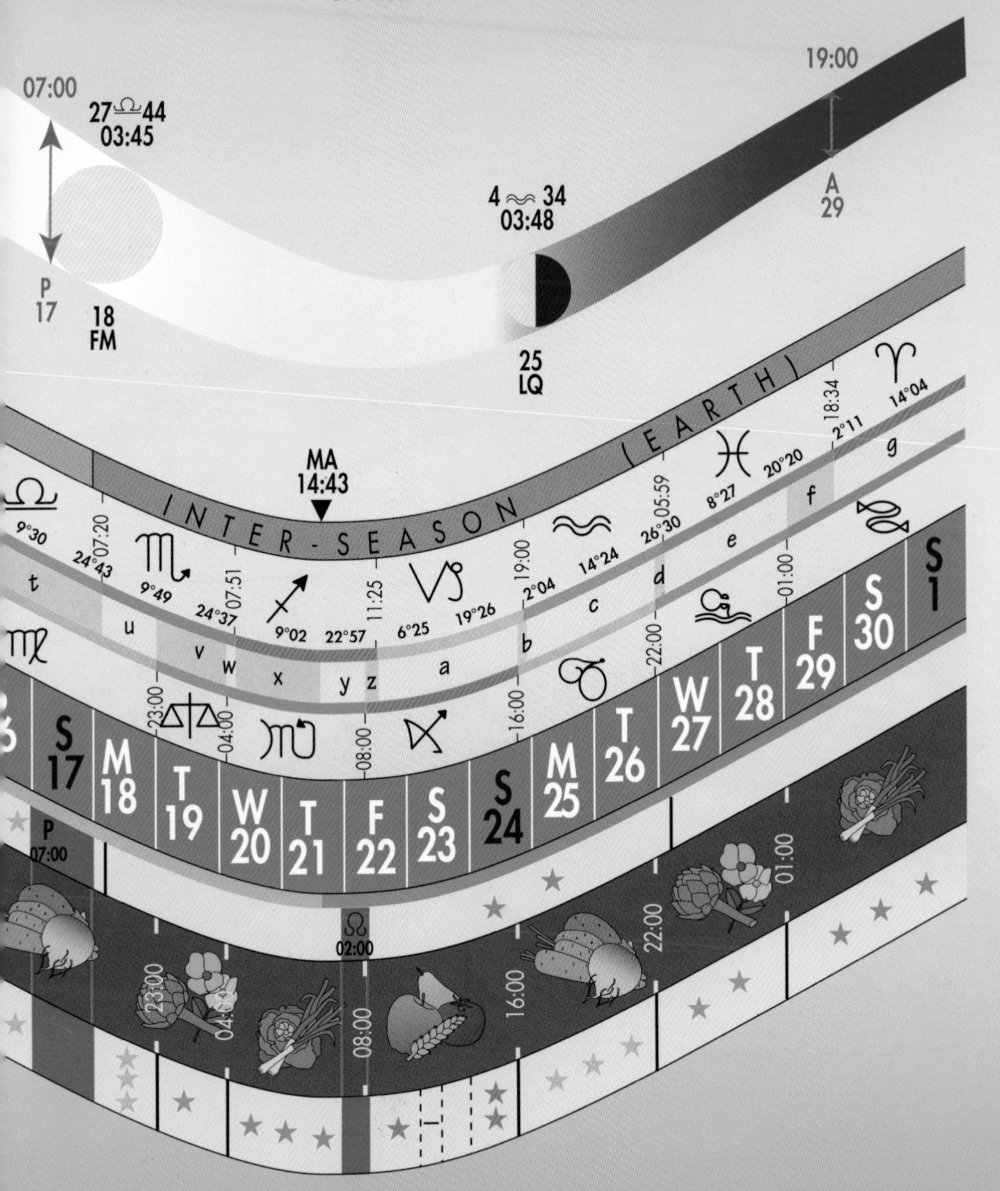

18-25-26 April are good days for planting potatoes.

MAY

3 May: the beginning of the Red Moon.

14 May: the Sun enters the constellation of Taurus.

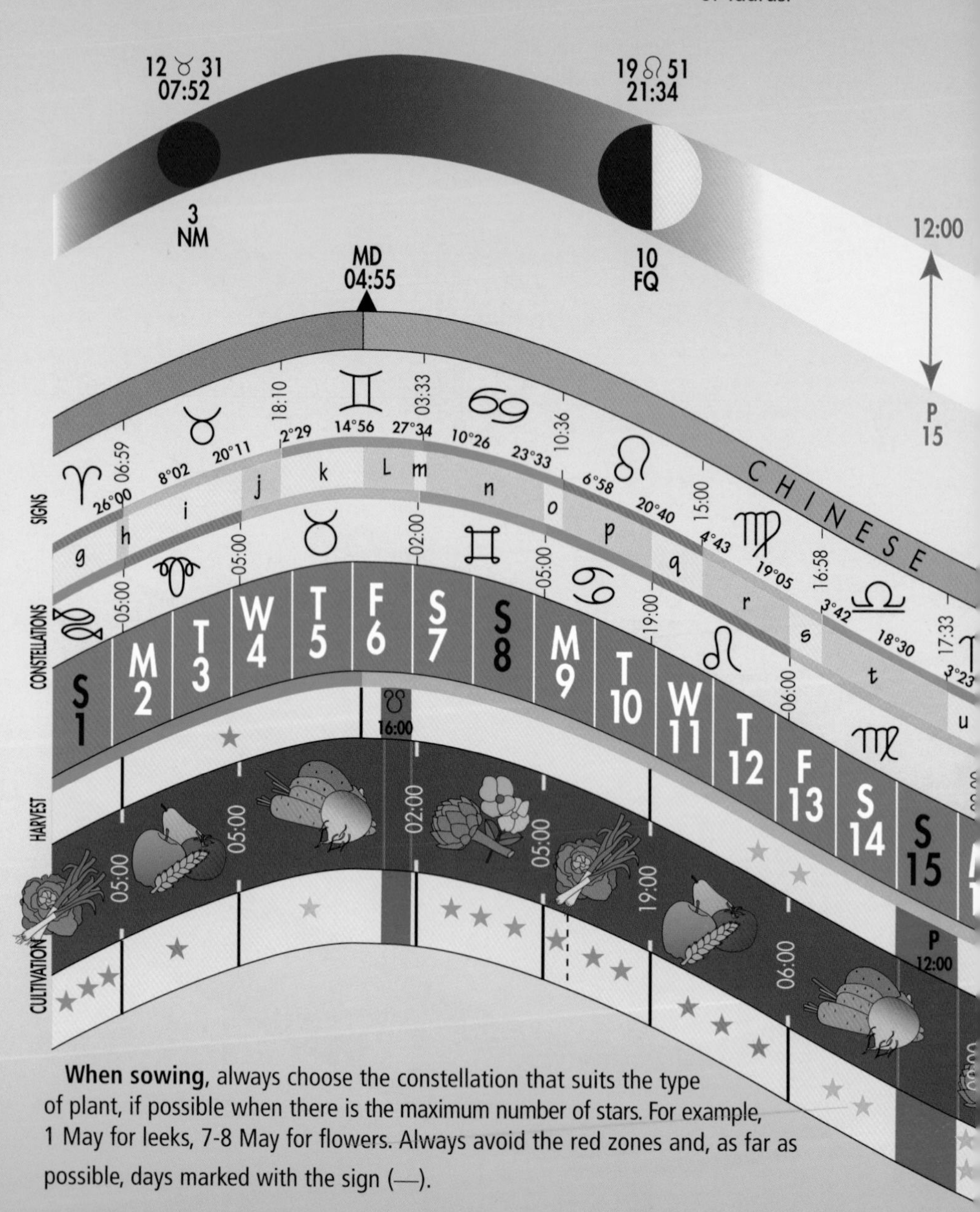

When sowing, always choose the constellation that suits the type of plant, if possible when there is the maximum number of stars. For example, 1 May for leeks, 7-8 May for flowers. Always avoid the red zones and, as far as possible, days marked with the sign (—).

Good days to **plant out flowers:** 7-8 May.

Note: all times given are in British Summer Time..

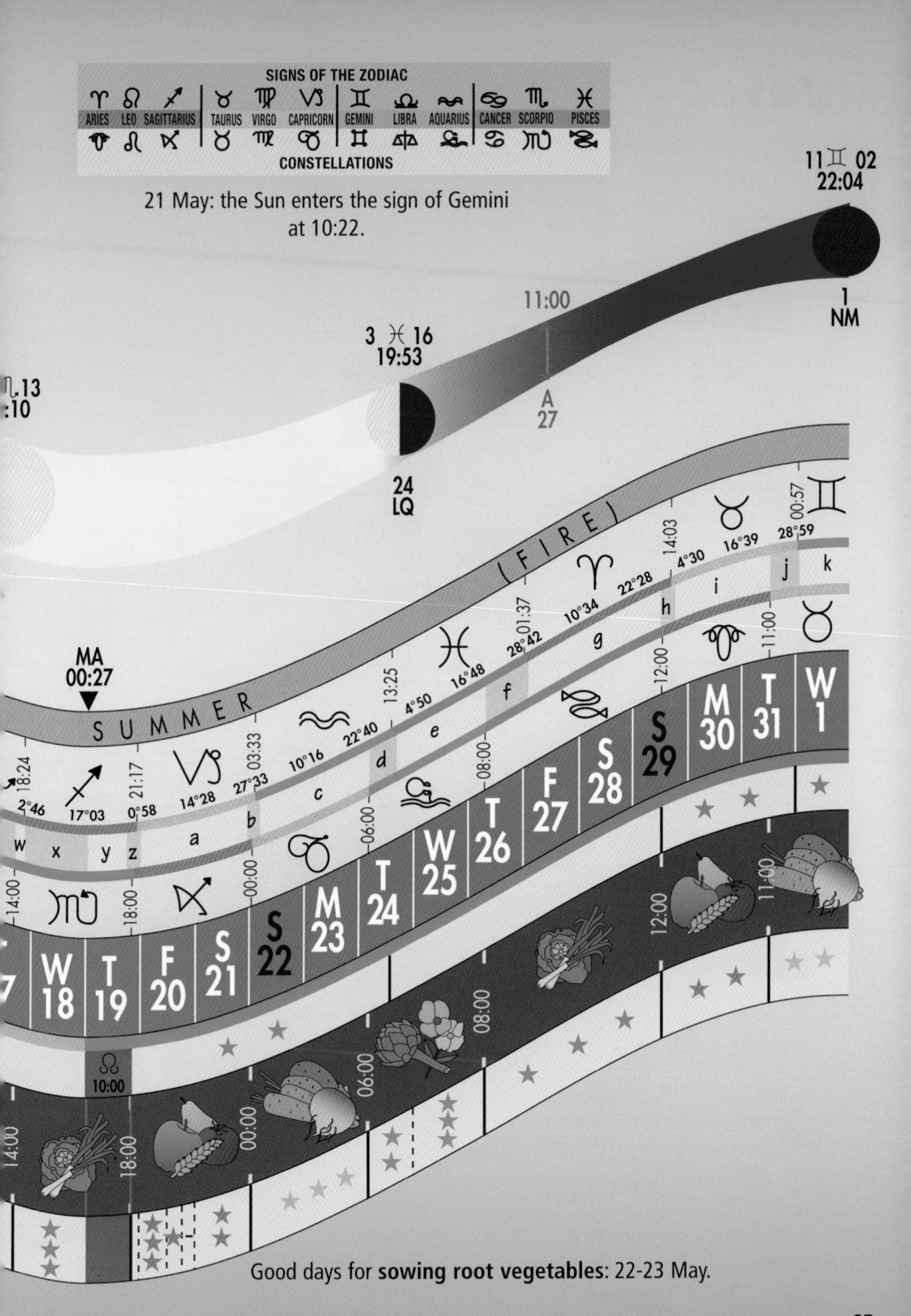

21 May: the Sun enters the sign of Gemini at 10:22.

Good days for **sowing root vegetables**: 22-23 May.

JUNE

1 June: end of the Red Moon.

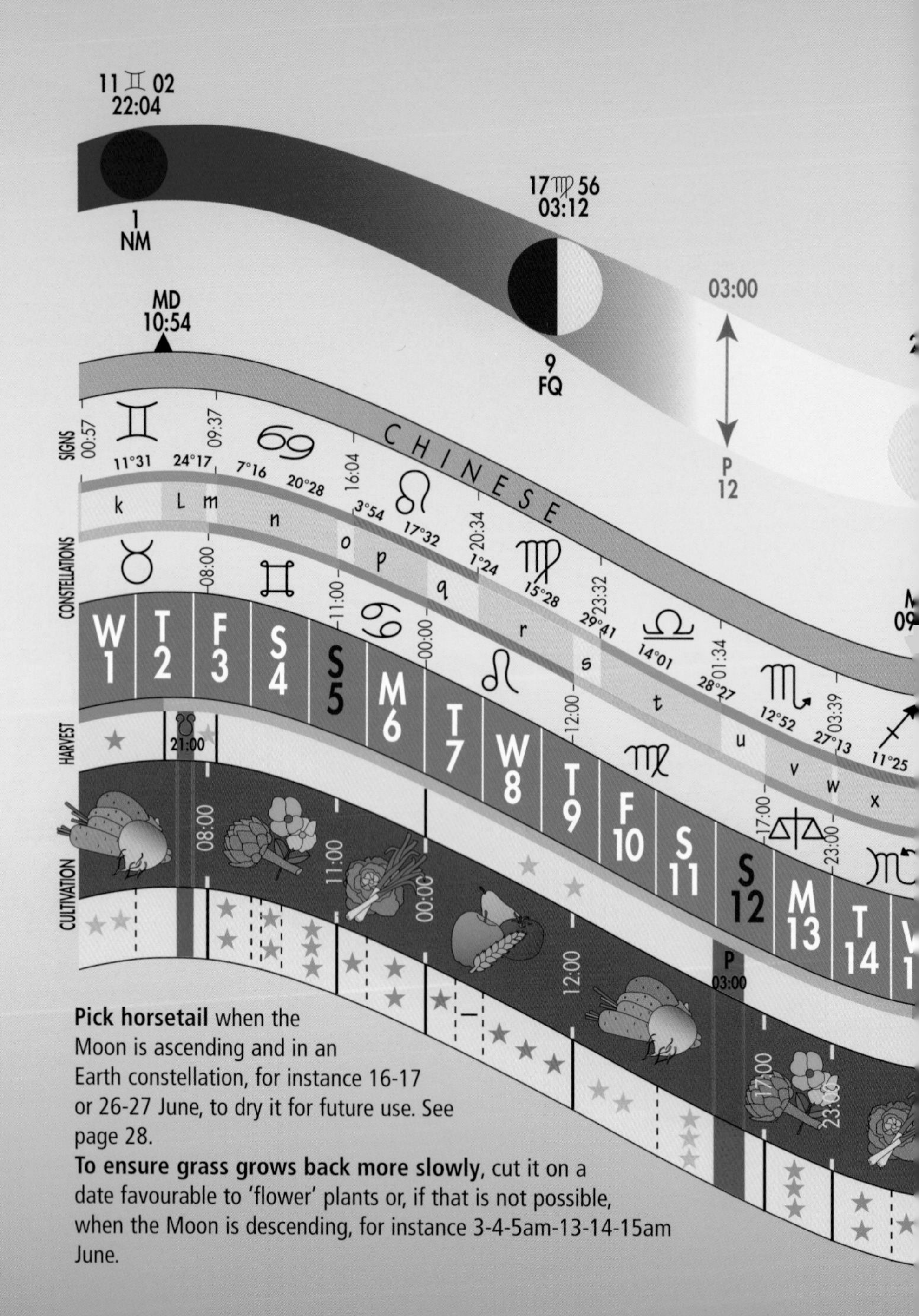

Pick horsetail when the Moon is ascending and in an Earth constellation, for instance 16-17 or 26-27 June, to dry it for future use. See page 28.

To ensure grass grows back more slowly, cut it on a date favourable to 'flower' plants or, if that is not possible, when the Moon is descending, for instance 3-4-5am-13-14-15am June.

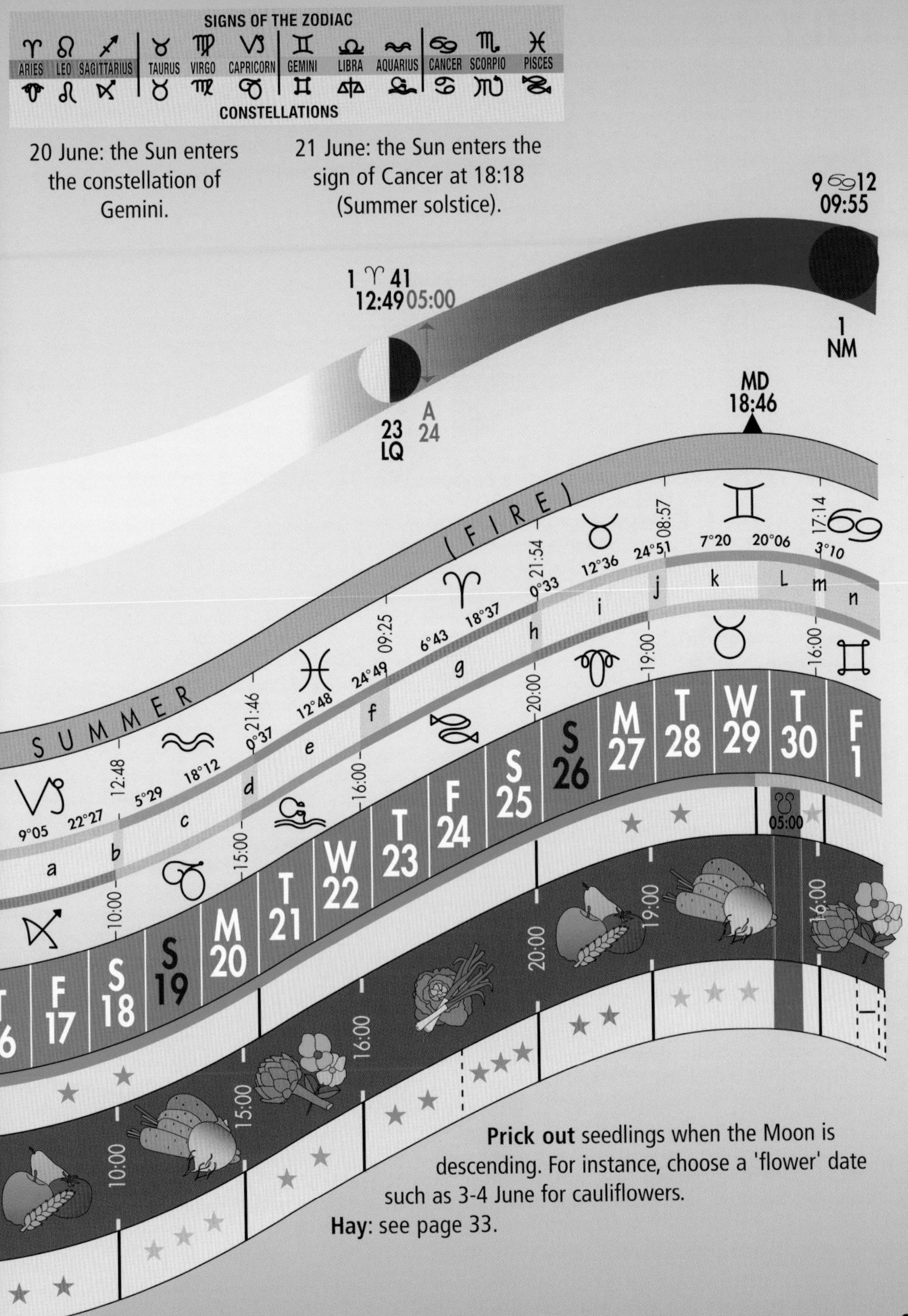

Prick out seedlings when the Moon is descending. For instance, choose a 'flower' date such as 3-4 June for cauliflowers.

Hay: see page 33.

JULY

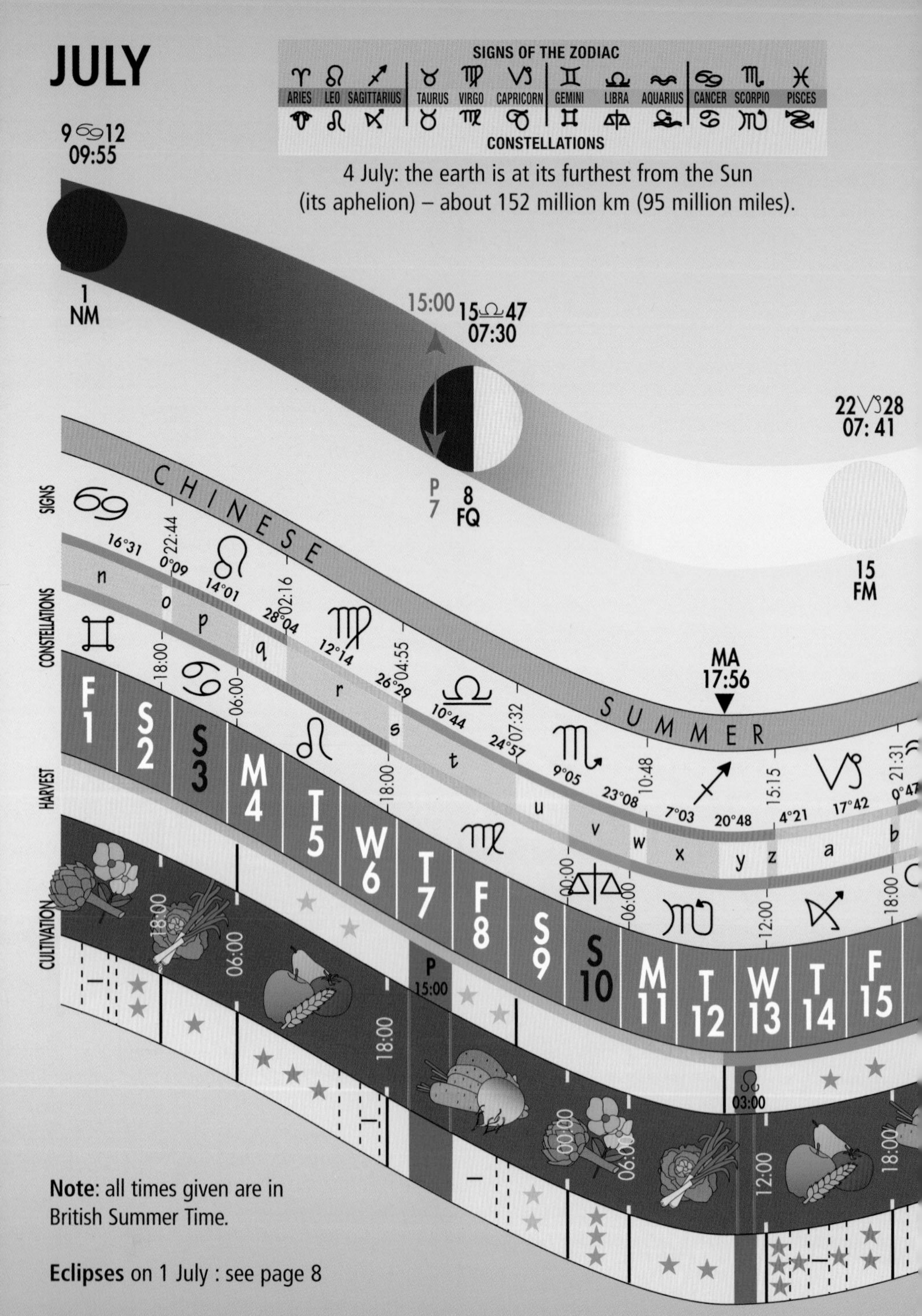

Note: all times given are in British Summer Time.

Eclipses on 1 July : see page 8

Good days to plant out strawberries: 4-5 July.

20 July: the Sun enters the constellation of Cancer.

23 July: the Sun enters the sign of Leo at 05:13.

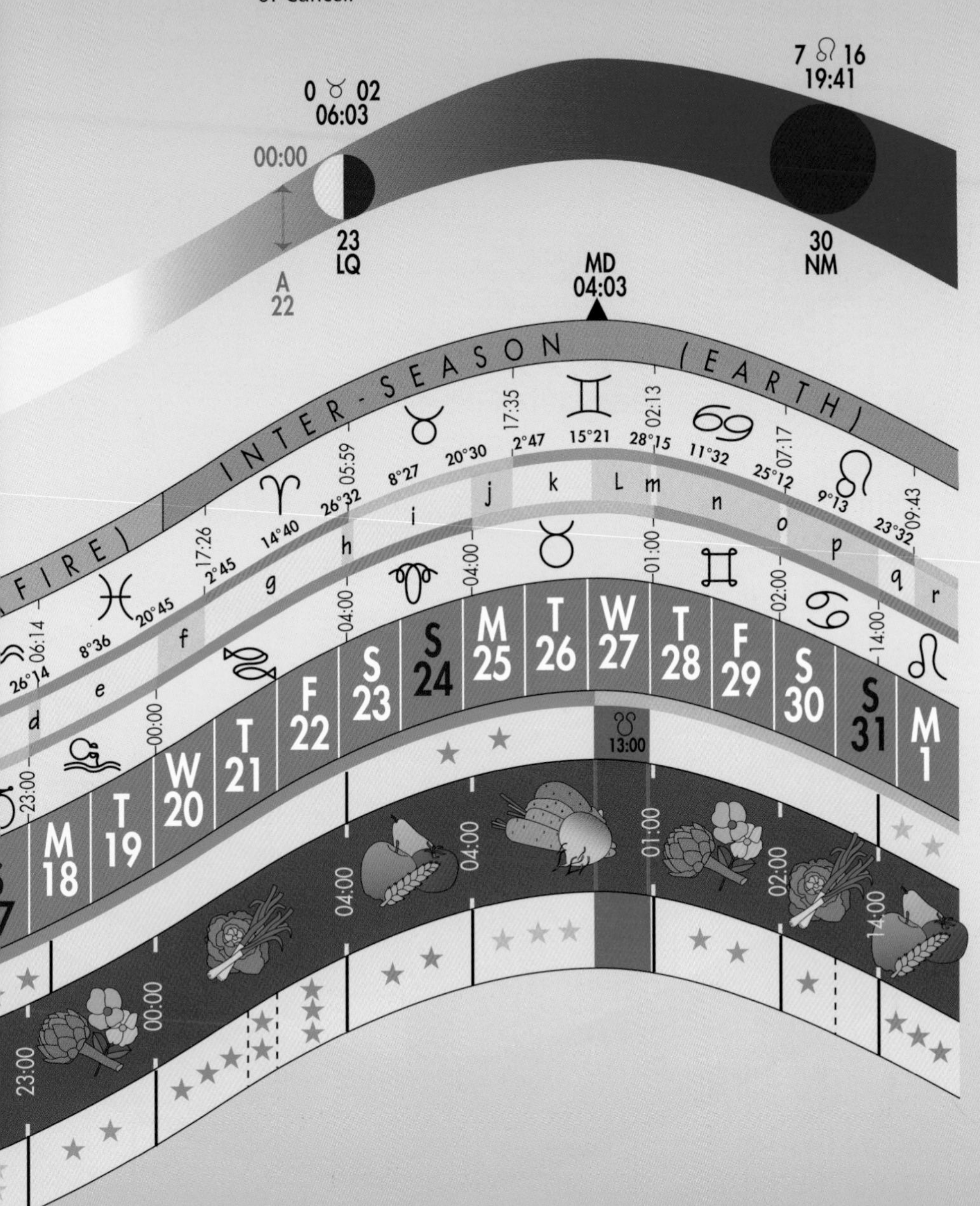

Water plants more during dry periods – to ensure roots grow deep, water generously but less often, rather than little and often. To avoid water evaporating too quickly, hoe the ground a few hours after watering, if possible in the evening with a waxing Moon in an Earth or Water sign, e.g. 1-2-5-6-9-10 July.

AUGUST

11 August: the Sun enters the constellation of Leo.

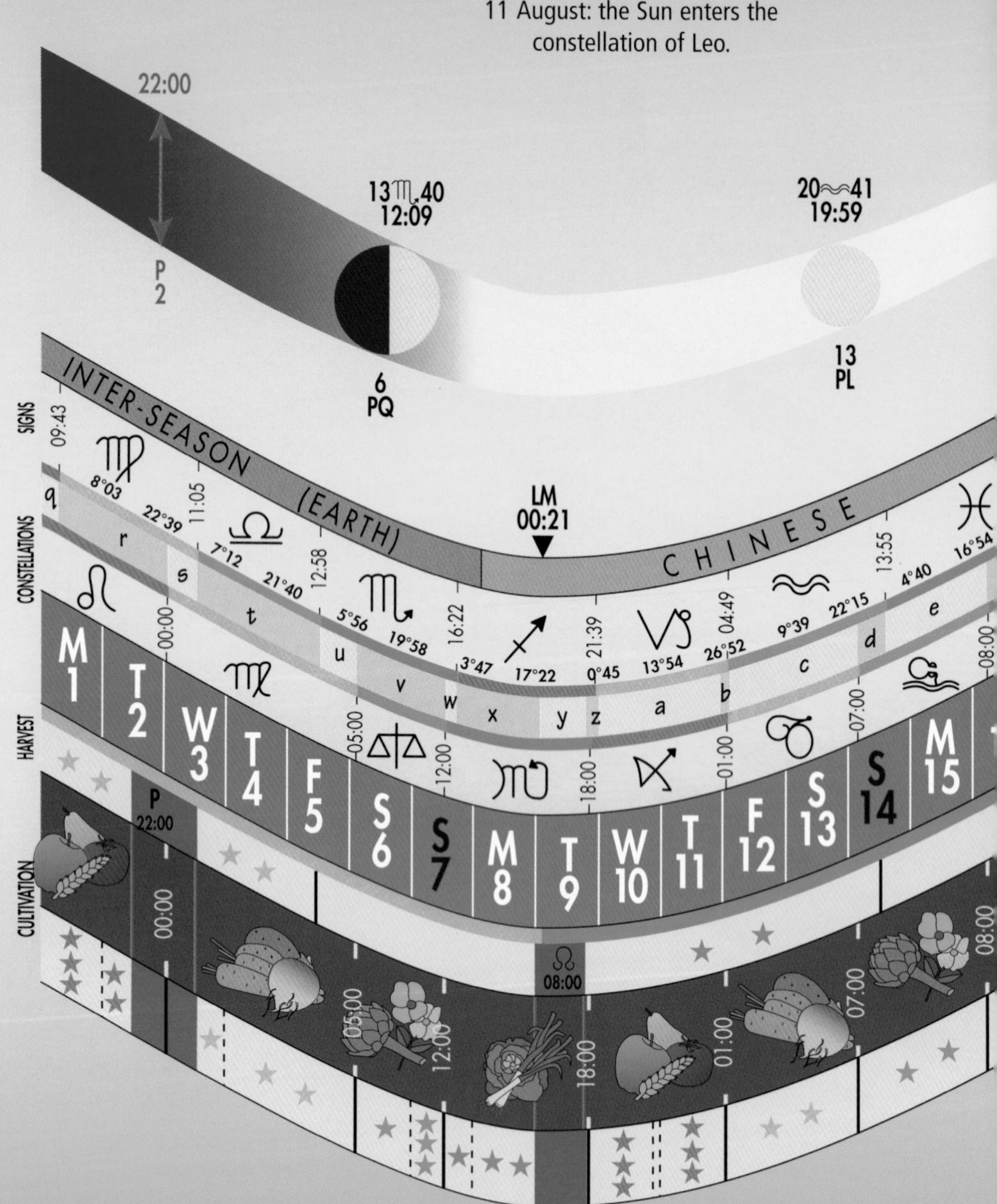

Shield budding: see page 26.

23 August: the Sun enters the sign of Virgo at 12:22.

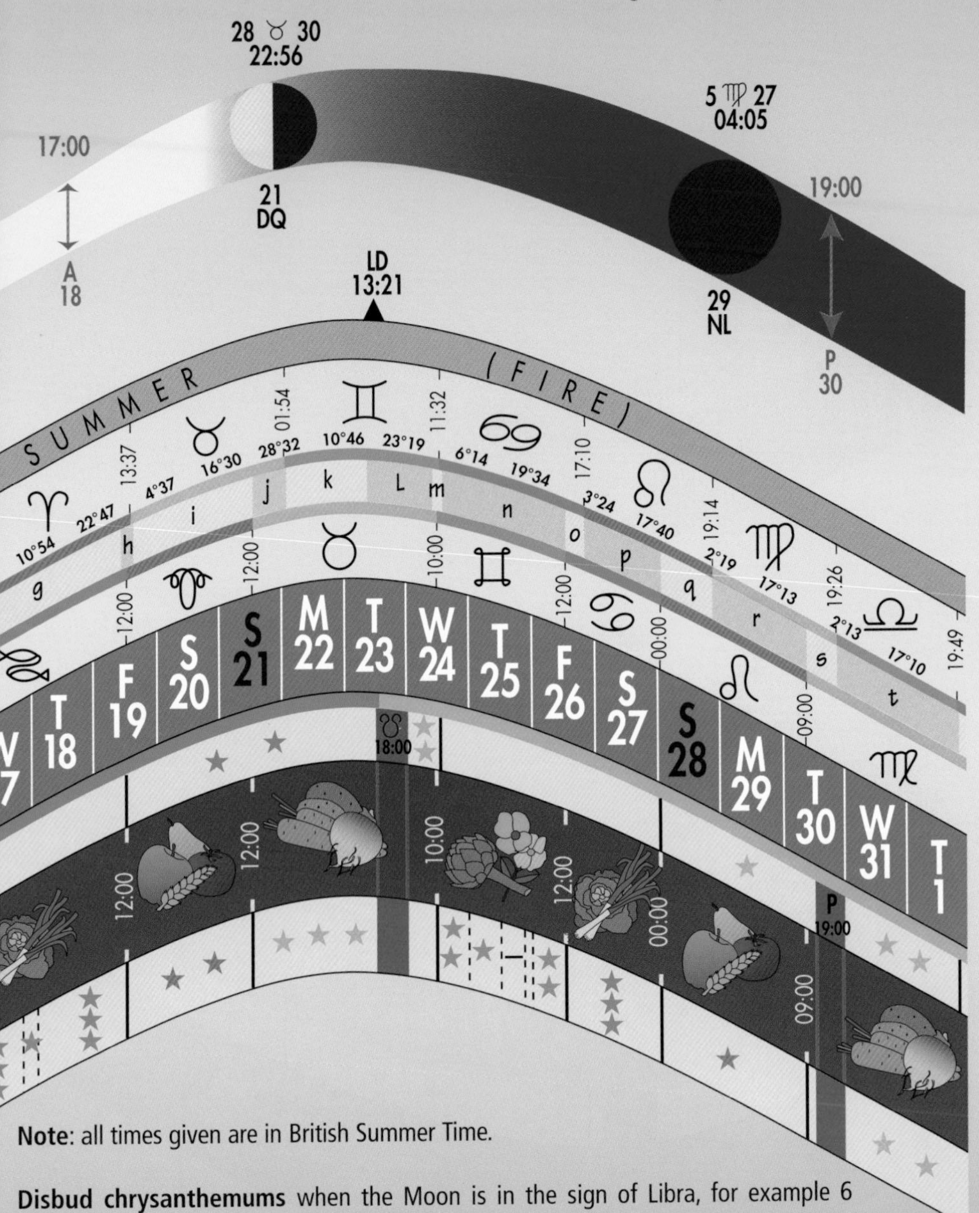

Note: all times given are in British Summer Time.

Disbud chrysanthemums when the Moon is in the sign of Libra, for example 6 August.

Take cuttings from geraniums: preferably when the Moon is ascending on a day favourable to 'flower' plants, for example, 14-15 August. Take the cutting from a healthy stem just below a leaf joint. Start the cutting off in a little peat or water; and when rooted, plant in soil-based compost.

SEPTEMBER

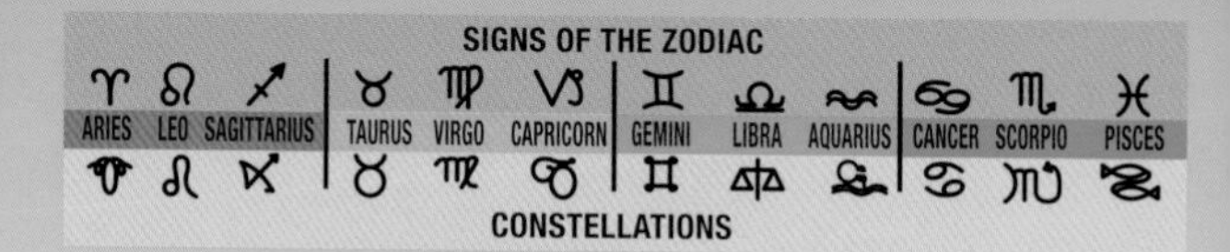

16 September: the Sun enters the constellation of Virgo.

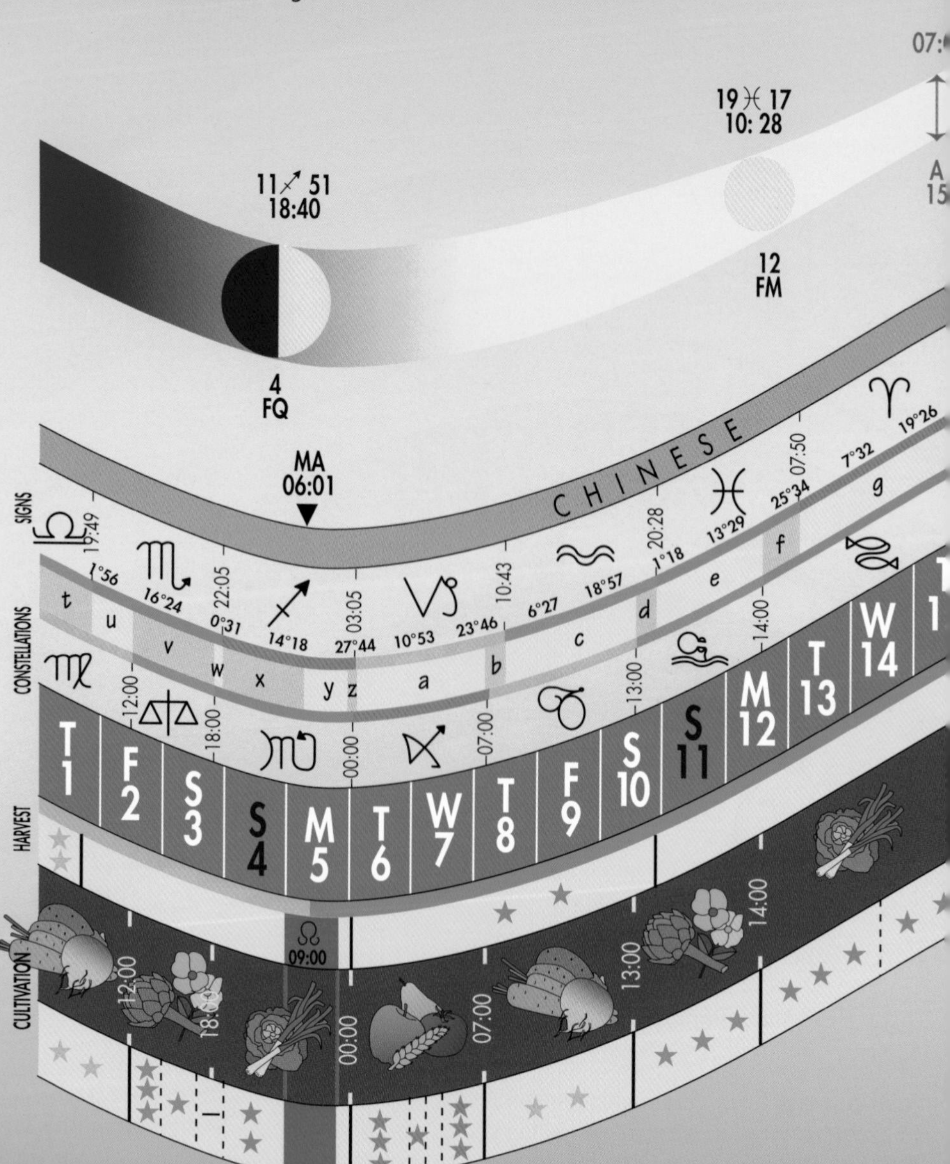

Harvesting grapes: see page 39.

Cereal crops: see page 35.

23 September: the Sun enters the sign of Libra at 10:06. (Autumnal equinox)

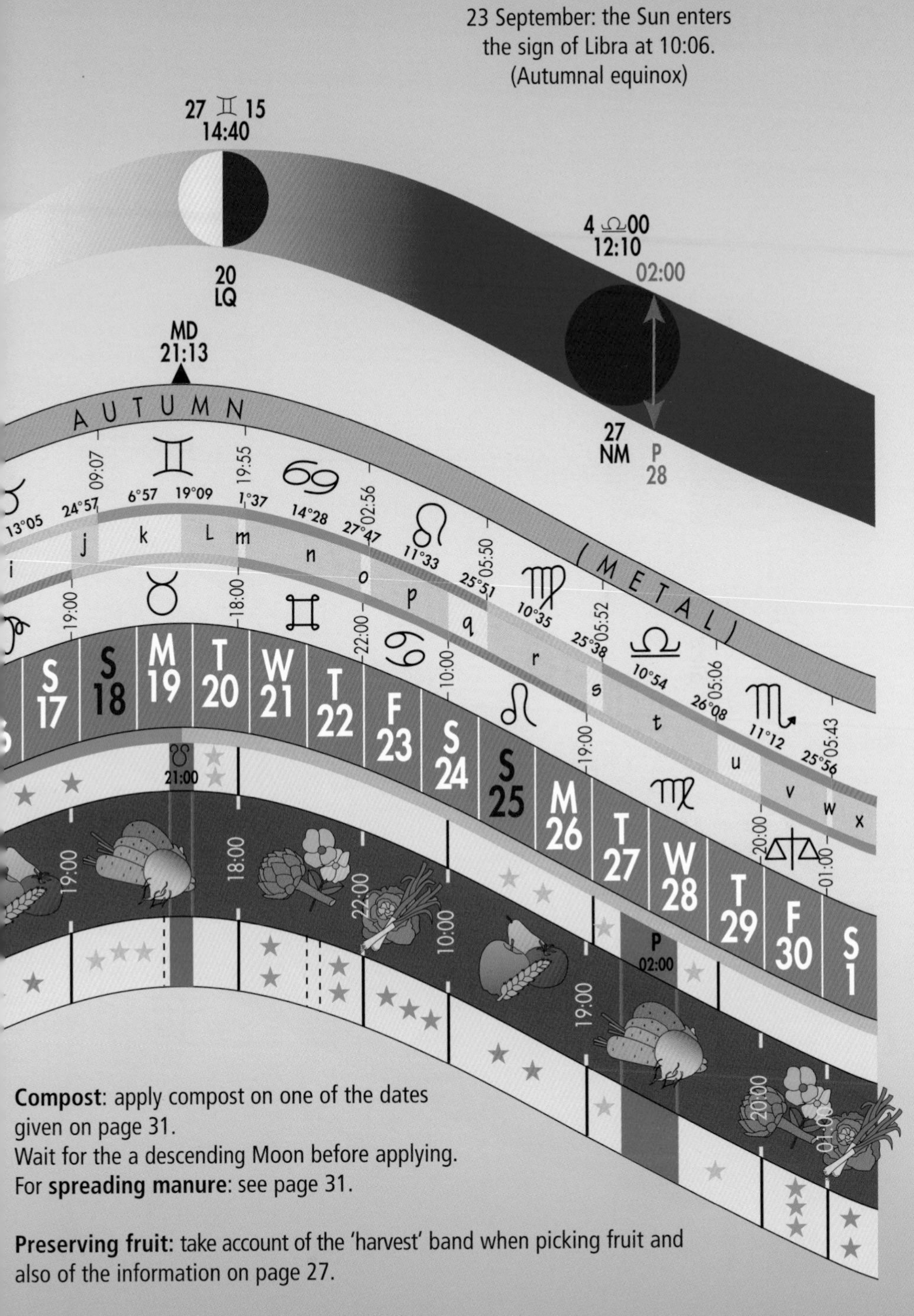

Compost: apply compost on one of the dates given on page 31.
Wait for the a descending Moon before applying.
For **spreading manure**: see page 31.

Preserving fruit: take account of the 'harvest' band when picking fruit and also of the information on page 27.

Note: all times given are in British Summer Time.

OCTOBER

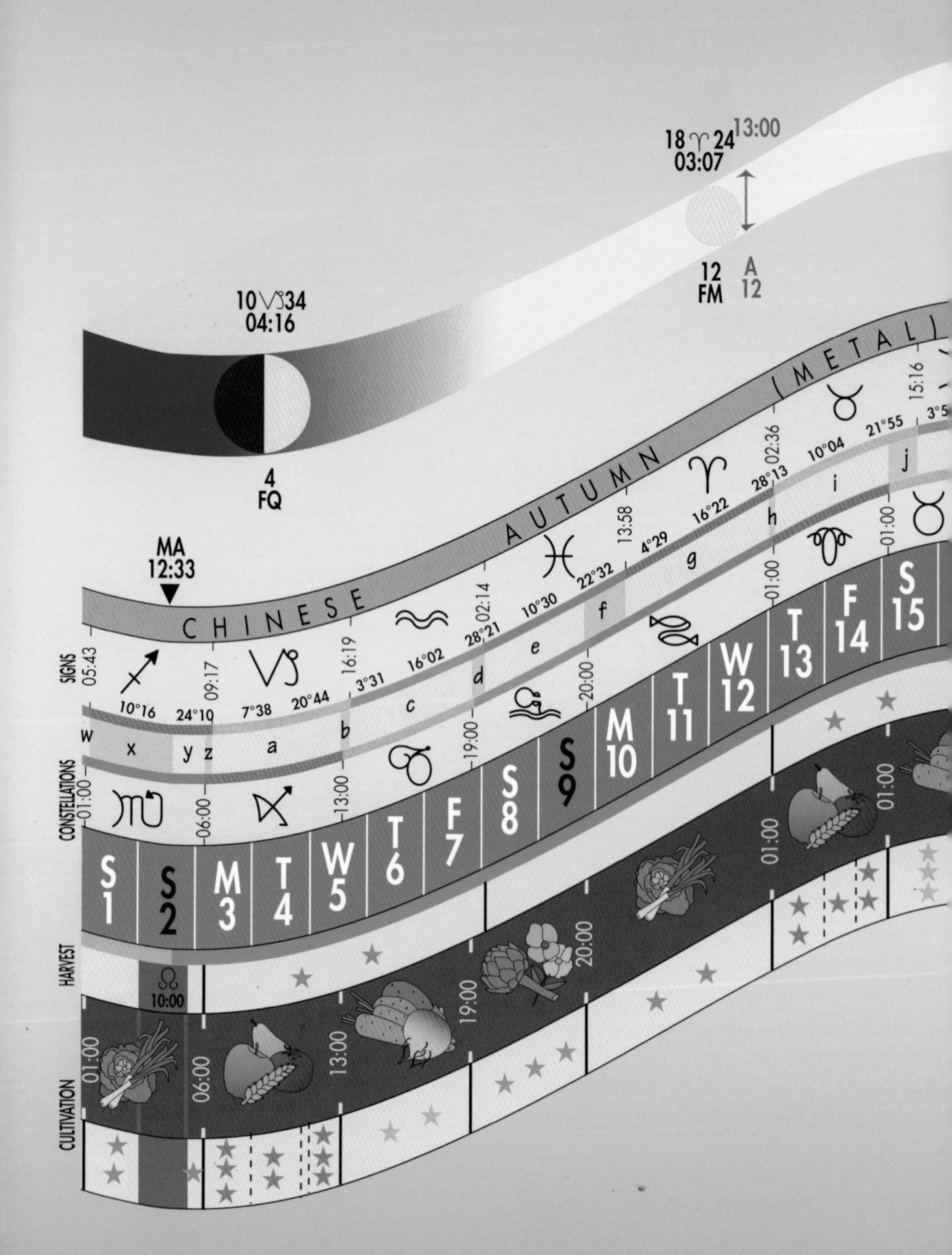

18 ♈ 24
03:07
13:00
12
FM
A
12
10 ♑ 34
04:16
4
FQ
MA
12:33
CHINESE AUTUMN (METAL)
SIGNS
CONSTELLATIONS
HARVEST
CULTIVATION
S 1
S 2
M 3
T 4
W 5
T 6
F 7
S 8
S 9
M 10
T 11
W 12
T 13
F 14
S 15

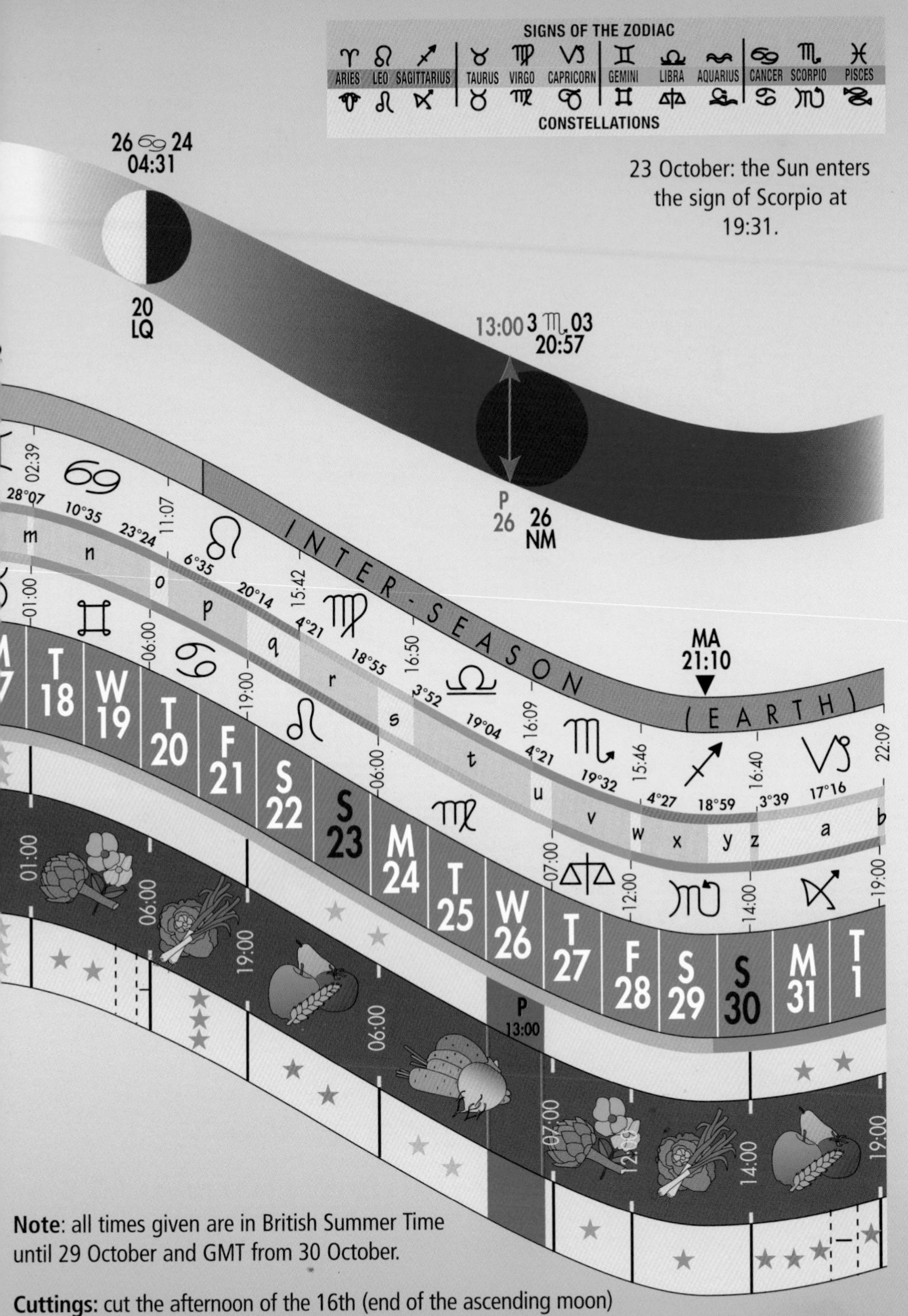

23 October: the Sun enters the sign of Scorpio at 19:31.

Note: all times given are in British Summer Time until 29 October and GMT from 30 October.

Cuttings: cut the afternoon of the 16th (end of the ascending moon) and plant during the morning of the 17th (beginning of the descending moon)

NOVEMBER

1 November: the Sun enters the constellation of Libra.

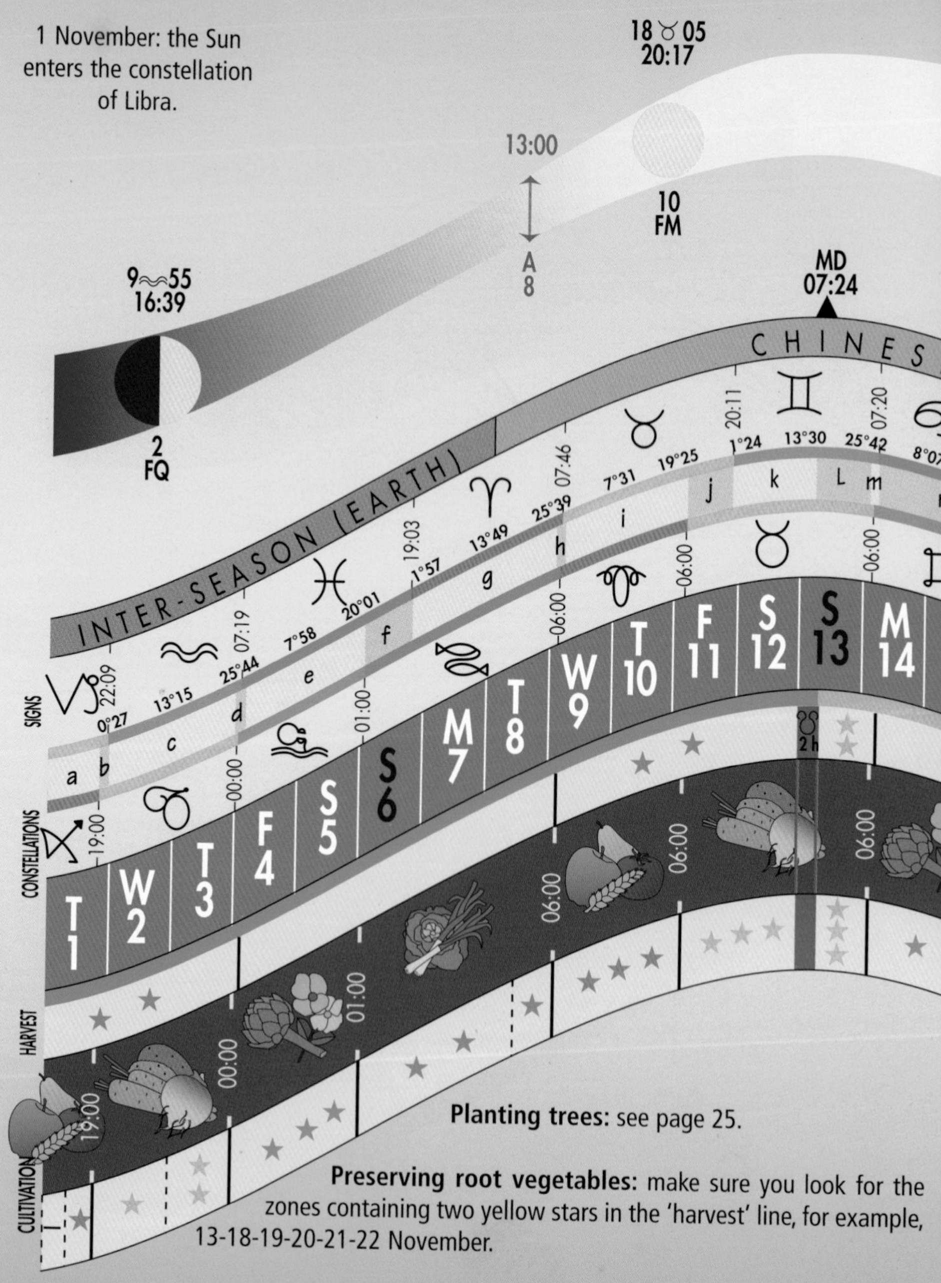

Planting trees: see page 25.

Preserving root vegetables: make sure you look for the zones containing two yellow stars in the 'harvest' line, for example, 13-18-19-20-21-22 November.

Timber: cut timber on the dates marked on page 39 to benefit from the lunar influences, if possible during November and December to benefit from the descending Sun as well.

Note: all times given are in Greenwich Mean Time.

DECEMBER

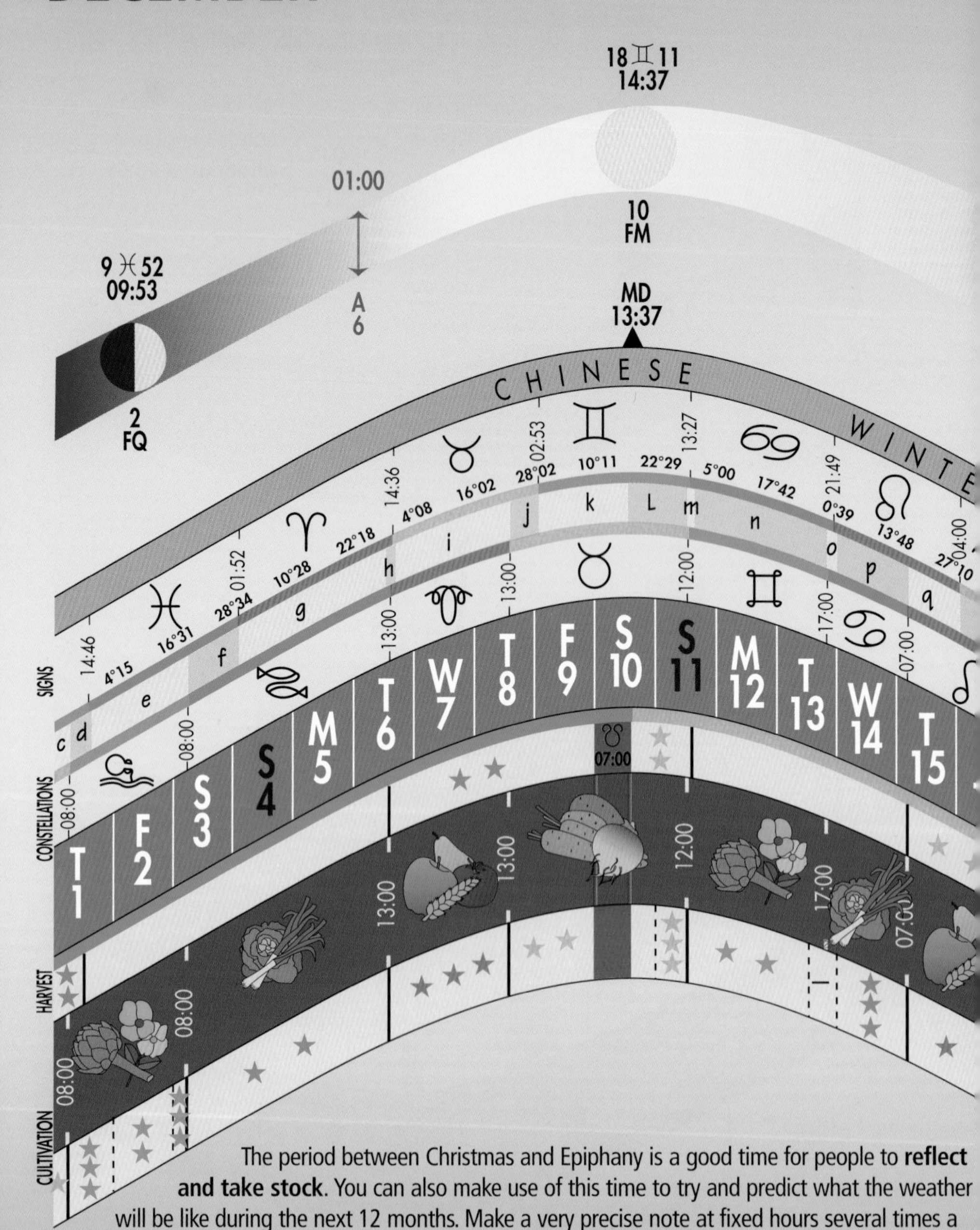

The period between Christmas and Epiphany is a good time for people to **reflect and take stock**. You can also make use of this time to try and predict what the weather will be like during the next 12 months. Make a very precise note at fixed hours several times a day of what the weather is like between 24 December and 6 January.

In theory, the weather in January will correspond to the weather as it was on 24/25 December, the weather in February to the weather conditions on 25/26 December and so on. Keep 24/25 December as your starting point. Try it and see if it works.

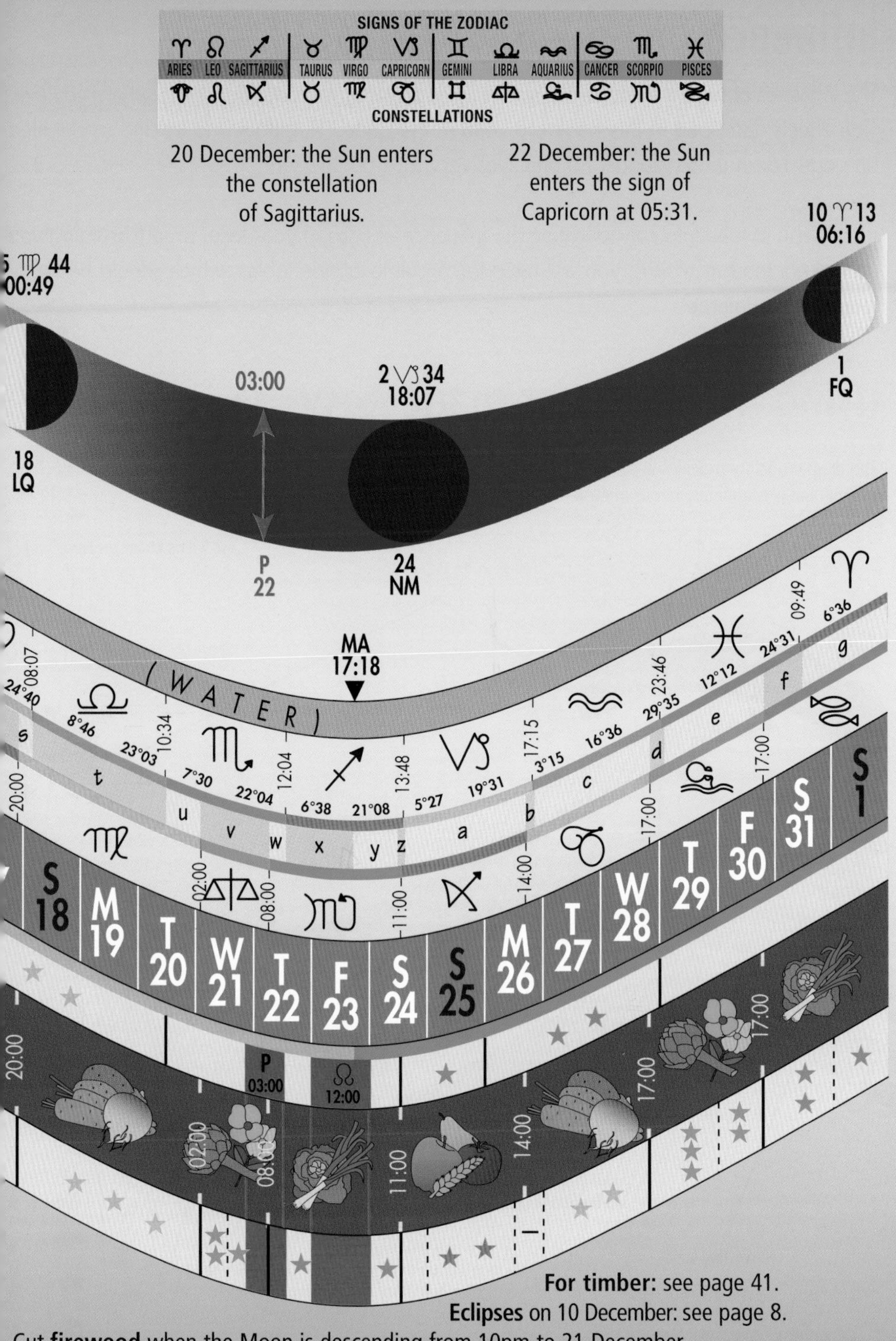

For timber: see page 41.
Eclipses on 10 December: see page 8.

Cut **firewood** when the Moon is descending from 10pm to 21 December.

Influence of tides... where you live

The influence of tides (see page 16) naturally depends on the tides nearest to where you live. Tide effects described in this book are valid in the United Kingdom and Ireland as detailed on page 16: in those islands, no one lives very far from the sea!

If you wish to take into consideration the influence of tides in your local area (assuming you do live not too far inland), you will need to consult local tide tables, which should be easily found on the internet.

Using the calendar made easy

The diagram below provides a clearer, simpler version of the information contained in the tear-out section. This information is a summary of our interpretation of the various lunar and planetary influences.

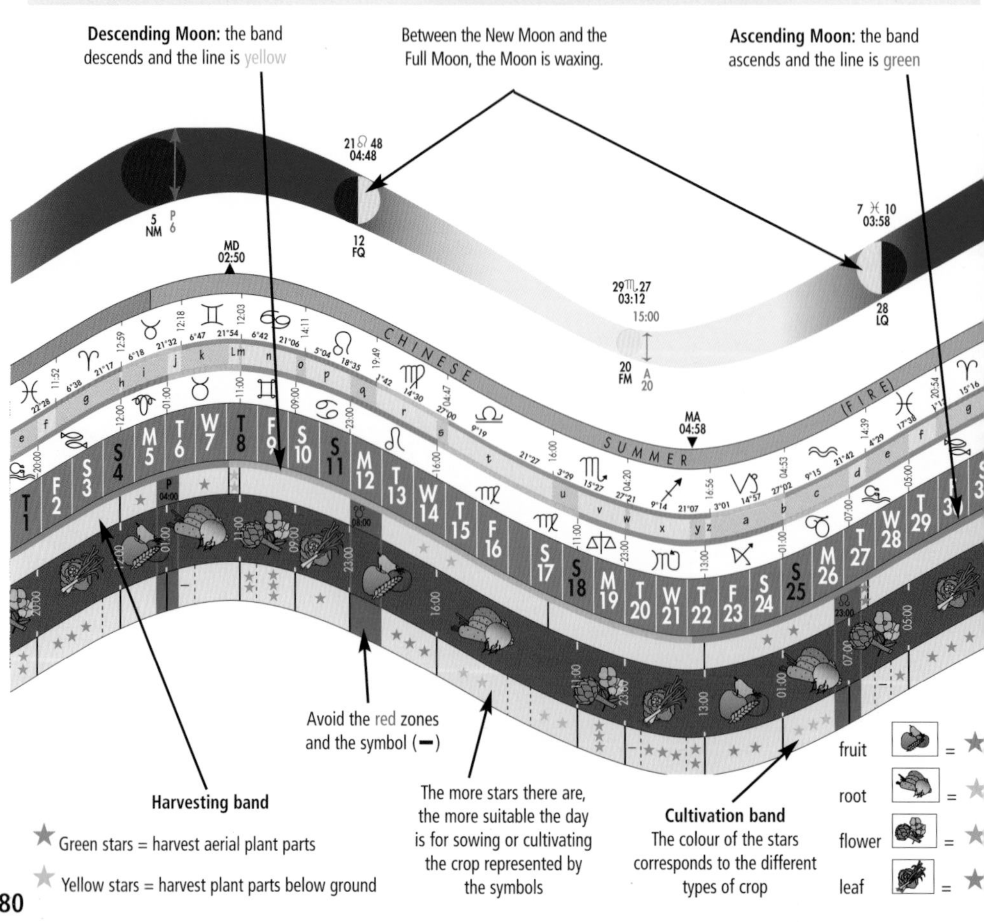

The note pages

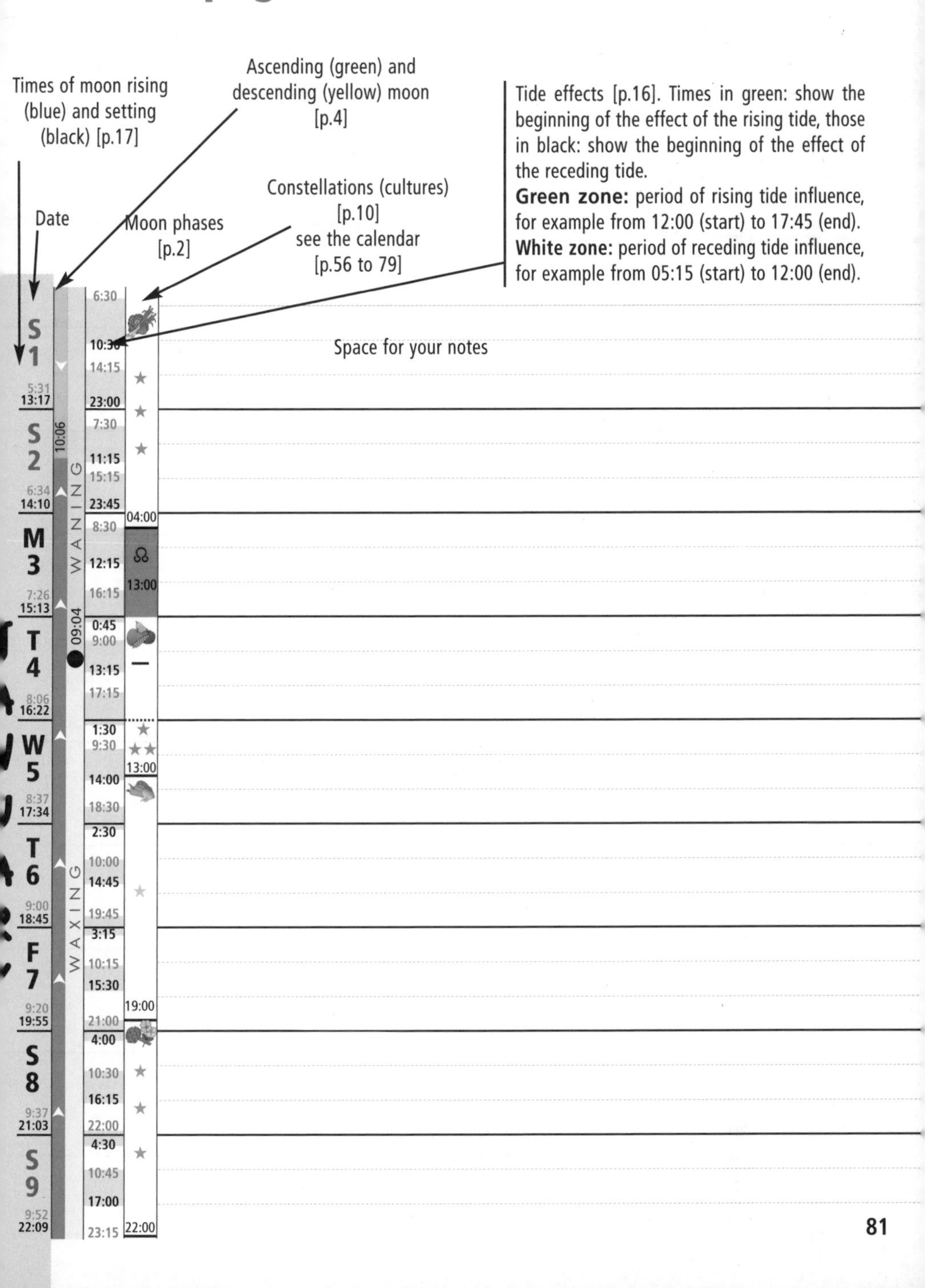

JANUARY
M 10
10:07
23:16
5:15
11:00
17:30
A
06:00
T 11
10:23
0:15
6:00
11:30
18:15
WAXING
11:33
W 12
0:23
10:40
1:30
6:30
11:45
19:00
02:00
T 13
1:32
11:01
2:30
7:15
12:00
19:45
F 14
2:41
11:28
3:45
8:15
12:30
20:30
00:00
S 15
3:50
12:02
4:45
9:00
13:00
21:30
WAXING
S 16
4:56
12:48
6:00
10:00
13:45
22:15
22:49
M 17
5:55
13:47
7:00
10:45
14:45
23:15
18:00
00:00
T 18
6:45
14:58
7:45
11:45
16:00
W 19
7:24
16:19
0:15
8:30
12:45
17:15
18:00
21:23
T 20
7:55
17:44
1:15
9:00
13:45
18:45
04:00
F 21
8:20
19:10
2:15
9:15
14:45
20:15
♉ ☿ 00:00
P
00:00
S 22
8:41
20:36
3:00
9:45
15:30
21:30
WANING

Day	Moonrise / Moonset	Times	Marks
S 23	9:01 / 22:00	4:00, 10:00, 16:15, 23:00	★ ★ 13:00
M 24	9:21 / 23:24	4:45, 10:15, 17:15	★ ★ ★
T 25	9:43	0:30, 5:45, 10:45, 18:00	
W 26	0:46 / 10:08	1:45, 6:30, 11:15, 19:00	★ 19:00 — ◑ 12:58
T 27	2:06 / 10:38	3:00, 7:30, 11:45, 20:00	★ ★ 02:00
F 28	3:21 / 11:17	4:15, 8:15, 12:15, 20:45	★
S 29	4:27 / 12:05	5:30, 9:15, 13:00, 21:45	★ ★ — 16:26
S 30	5:22 / 13:03	6:15, 10:15, 14:00, 22:45	★ 11:00 ☊ 18:00
M 31	6:05 / 14:09	7:00, 11:00, 15:15, 23:30	★
T 1	6:39 / 15:19	7:45, 12:00, 16:15	★ 20:00
W 2	7:05 / 16:30	0:15, 8:00, 12:45, 17:30	★ ★ — ● 02:32
T 3	7:25 / 17:40	1:00, 8:30, 13:30, 18:45	03:00
F 4	7:43 / 18:49	1:45, 8:45, 14:15, 19:45	★ ★ ★

WANING

WANING

WAXING

FEBRUARY
S 5 7:59 19:56 2:30 9:00 14:45 21:00 05:00
S 6 8:14 21:02 3:15 9:15 15:30 22:00 23:00 A
M 7 8:30 22:09 3:45 9:30 16:15 23:15
T 8 8:46 23:16 4:30 9:45 17:00
W 9 9:06 0:15 5:15 10:00 17:45 10:00
T 10 0:24 9:30 1:30 6:00 10:30 18:30
F 11 1:32 10:00 2:30 6:45 11:00 19:15 09:00
07:19
S 12 2:38 10:39 3:45 7:45 11:45 20:00
S 13 3:39 11:29 4:45 8:30 12:30 21:00
09:00
M 14 4:32 12:33 5:30 9:30 13:30 22:00 04:00 08:00
T 15 5:16 13:48 6:15 10:30 14:45 23:00 04:00
W 16 5:51 15:10 6:45 11:30 16:15
T 17 6:19 16:37 0:00 7:15 12:15 17:30 15:00
WAXING
WAXING

Day	Moon	Times	Notes
F 18	6:43 / 18:05	0:45, 7:45, 13:15, 19:00	08:37
S 19	7:04 / 19:32	1:45, 8:00, 14:15, 20:30	P 07:00 – 22:00
S 20	7:25 / 20:59	2:30, 8:30, 15:00, 22:00	
M 21	7:47 / 22:26	3:30, 8:45, 16:00, 23:30	
T 22	8:11 / 23:50	4:30, 9:15, 16:45	02:00
W 23	8:41	0:45, 5:15, 9:45, 17:45	
T 24	1:09 / 9:17	2:15, 6:15, 10:15, 18:45	08:00; 23:27
F 25	2:19 / 10:03	3:15, 7:15, 11:00, 19:45	22:12
S 26	3:18 / 10:59	4:15, 8:15, 12:00, 20:30	16:00; ☊ 20:00
S 27	4:05 / 12:02	5:00, 9:00, 13:00, 21:30	
M 28	4:41 / 13:11	5:45, 10:00, 14:15, 22:15	02:00
T 1	5:09 / 14:20	6:15, 10:45, 15:15, 23:00	
W 2	5:31 / 15:30	6:30, 11:30, 16:30, 23:45	

WANING

WANING

MARCH
T 3
F 4
S 5
S 6
M 7
T 8
W 9
T 10
F 11
S 12
S 13
M 14
T 15
WANING
WAXING
20:47
17:08
23:46
09:00
11:00
08:00
16:00
16:00
12:00
13:00
15:00
☊☿ 17:00
♉♀ 08:00

Day	Rise / Set	Moon	Times	Notes
W 16	4:16 / 14:04	WAXING	5:15, 10:00, 15:00, 22:30	★ ★ 02:00
T 17	4:41 / 15:29		5:45, 11:00, 16:30, 23:30	★
F 18	5:04 / 16:56		6:00, 11:45, 18:00	★ ★
S 19	5:25 / 18:24	○ 18:11	0:15, 6:30, 12:45, 19:30	★ ★ 10:00; P 19:00
S 20	5:47 / 15:29		1:15, 6:45, 13:45, 21:00	★
M 21	6:11 / 21:21		2:15, 7:15, 14:30, 22:15	★ ★
T 22	22:46 / 6:40	WANING	3:00, 7:45, 15:30, 23:45	11:00
W 23	7:15		4:00, 8:15, 16:30	★ ★ 16:00
T 24	0:03 / 7:59	05:04	1:00, 5:00, 9:00, 17:30	★ ★ ★
F 25	1:09 / 8:52		2:15, 6:00, 9:45, 18:30	★ ★ 21:00 ☊
S 26	2:01 / 9:55	◑ 12:08	3:00, 7:00, 11:00, 19:30	23:00 ★
S 27	3:41 / 12:03	WANING	4:45, 8:45, 13:00, 21:15	★ ★
M 28	4:12 / 13:12		5:15, 9:45, 14:15, 22:00	09:00 ★ ★ ★

MARCH
T 29
W 30
T 31
APRIL
F 1
S 2
S 3
M 4
T 5
W 6
T 7
F 8
S 9
S 10
WANING
WAXING
15:33
23:57
16:00
18:00
10:00
A
23:00
23:00
14:00
20:00

APRIL
S 24
3:15
7:30
12:00
16:00
2:12
11:00
20:00
03:48
M 25
3:45
8:15
13:15
2:39
12:11
20:45
T 26
4:00
9:00
14:15
3:00
13:20
21:30
22:00
W 27
4:15
9:45
15:30
3:19
14:27
22:15
T 28
4:30
10:30
16:30
3:35
15:34
22:45
01:00
WANING
F 29
4:45
11:15
17:45
A
3:51
16:40
23:30
19:00
S 30
5:15
11:45
4:08
17:46
18:45
MAY
S 1
0:15
5:30
12:30
4:25
18:53
20:00
1:00
05:00
M 2
5:45
13:15
4:46
20:00
21:00
07:52
T 3
1:45
6:15
14:00
5:10
21:07
22:00
W 4
2:30
05:00
6:45
14:45
5:41
22:12
23:15
T 5
3:15
7:15
15:45
WAXING
6:20
23:10
04:55
F 6
0:15
4:15
8:15
16:30
16:00
7:09

MAY

Day	Moonrise/set	Times	Notes
S 7	0:01 8:08	1:00 5:00 9:15 17:30	02:00
S 8	0:43 9:16	1:45 6:00 10:15 18:30	
M 9	1:16 10:30	2:15 6:45 11:30 19:15	05:00
T 10	1:44 11:47	2:45 7:45 12:45 20:00	21:34 19:00
W 11	2:08 13:07	3:15 8:30 14:00 21:00	
T 12	2:29 14:28	3:30 9:15 15:30 21:45	
F 13	2:49 15:51	3:45 10:15 16:45 22:45	06:00
S 14	3:10 17:15	4:15 11:00 18:15 23:30	
S 15	3:34 18:41	4:30 12:00 19:45	P 12:00
M 16	4:03 20:06	0:30 5:00 13:00 21:00	09:00
T 17	4:39 21:25	1:30 5:45 14:00 22:30	12:10 14:00
W 18	5:25 22:33	2:30 6:30 15:00 23:30	00:27
T 19	6:22 23:27	3:30 7:15 16:00	☊ 10:00 18:00

WAXING

WAXING

WANING

MAY
F 20
7:29
0:30
4:30
8:30
17:00
S 21
0:08
8:41
1:15
5:15
9:45
17:45
00:00
S 22
0:39
9:54
1:45
6:15
11:00
18:30
M 23
1:03
11:05
2:00
7:00
12:00
19:15
T 24
1:23
12:14
2:30
7:45
13:15
20:00
06:00
19:53
W 25
1:41
13:22
2:45
8:30
14:15
20:45
T 26
1:57
14:28
3:00
9:00
15:30
21:30
08:00
F 27
2:13
15:34
3:15
9:45
16:30
22:15
A
11:00
S 28
2:30
16:41
3:30
10:30
17:45
22:45
S 29
2:50
17:48
3:45
11:15
18:45
23:30
12:00
M 30
3:13
18:56
4:15
12:00
20:00
T 31
3:41
20:02
0:15
4:45
12:45
21:00
11:00
W 1
4:18
21:03
1:15
5:15
13:30
22:00
22:04
WANING
WANING

JUNE

WAXING

Day	Times	Times	Symbols
T 2	5:04 21:57	2:00 6:00 14:30 23:00	10:54 ♉ 21:00
F 3	6:00 22:42	3:00 7:00 15:30 23:45	08:00 ★ ★
S 4	7:07 23:19	3:45 8:00 16:15	★ ★ ★ ★
S 5	8:20 23:48	0:15 4:45 9:15 17:15	★ 11:00 ★
M 6	9:37	0:45 5:45 10:30 18:00	★ ★ 00:00
T 7	0:13 10:55	1:15 6:30 12:00 19:00	★ — ☊☿ 17:00
W 8	0:34 12:15	1:30 7:15 13:15 19:45	★ ★ ★ ◐ 03:12
T 9	0:55 13:35	2:00 8:00 14:30 20:30	12:00
F 10	1:15 14:57	2:15 9:00 16:00 21:30	★ ★
S 11	1:37 16:19	2:30 9:45 17:15 22:15	★ ★ ★ P 03:00
S 12	2:02 17:42	3:00 10:45 18:45 23:15	17:00
M 13	2:34 19:02	3:30 11:45 20:00	★ ★ ★ 23:00
T 14	3:14 20:14	0:15 4:15 12:45 21:15	★ ★

JUNE
W 15
4:05
21:14
09:53
21:15
1:15
5:00
13:45
22:15
20:00
04:00
T 16
5:08
22:01
2:15
6:15
14:45
23:00
F 17
6:18
22:37
3:00
7:15
15:30
23:30
S 18
7:32
23:05
4:00
8:30
16:30
10:00
S 19
8:46
23:27
0:00
4:45
9:45
17:15
M 20
9:57
23:45
0:30
5:30
11:00
18:00
15:00
T 21
11:06
0:45
6:15
12:00
18:45
WANING
W 22
0:02
12:14
1:00
7:00
13:15
19:15
16:00
T 23
0:19
13:20
12:49
1:15
7:45
14:15
20:00
F 24
0:35
14:26
1:30
8:30
15:30
20:45
A
05:00
S 25
0:54
15:33
2:00
9:00
16:30
21:30
20:00
WANING
S 26
1:15
16:41
2:15
9:45
17:45
22:15
M 27
1:41
17:47
2:45
10:45
18:45
23:00
19:00

JULY
T 28
2:14
18:51
3:15
11:30
19:45
W 29
2:56
19:49
0:00
4:00
12:15
20:45
18:46
T 30
3:49
20:38
0:45
4:45
13:15
21:45
05:00
♉
16:00
WANING
09:55
F 1
4:52
21:19
1:45
5:45
14:15
22:15
S 2
6:05
21:51
2:45
7:00
15:00
22:45
18:00
S 3
7:23
22:18
3:30
8:30
16:00
23:15
06:00
M 4
8:42
22:41
4:30
9:45
16:45
23:45
T 5
10:03
23:01
5:15
11:00
17:45
W 6
11:24
23:22
0:00
6:00
12:30
18:30
☊ ♀ 12:00
18:00
T 7
12:45
23:43
0:15
7:00
13:45
19:15
P
15:00
WAXING
F 8
14:06
07:30
0:45
7:45
15:00
20:15
☊ ♂ 10:00
S 9
0:07
15:27
1:00
8:45
16:30
21:00
00:00
S 10
0:35
16:47
1:30
9:30
17:45
22:00
WAXING

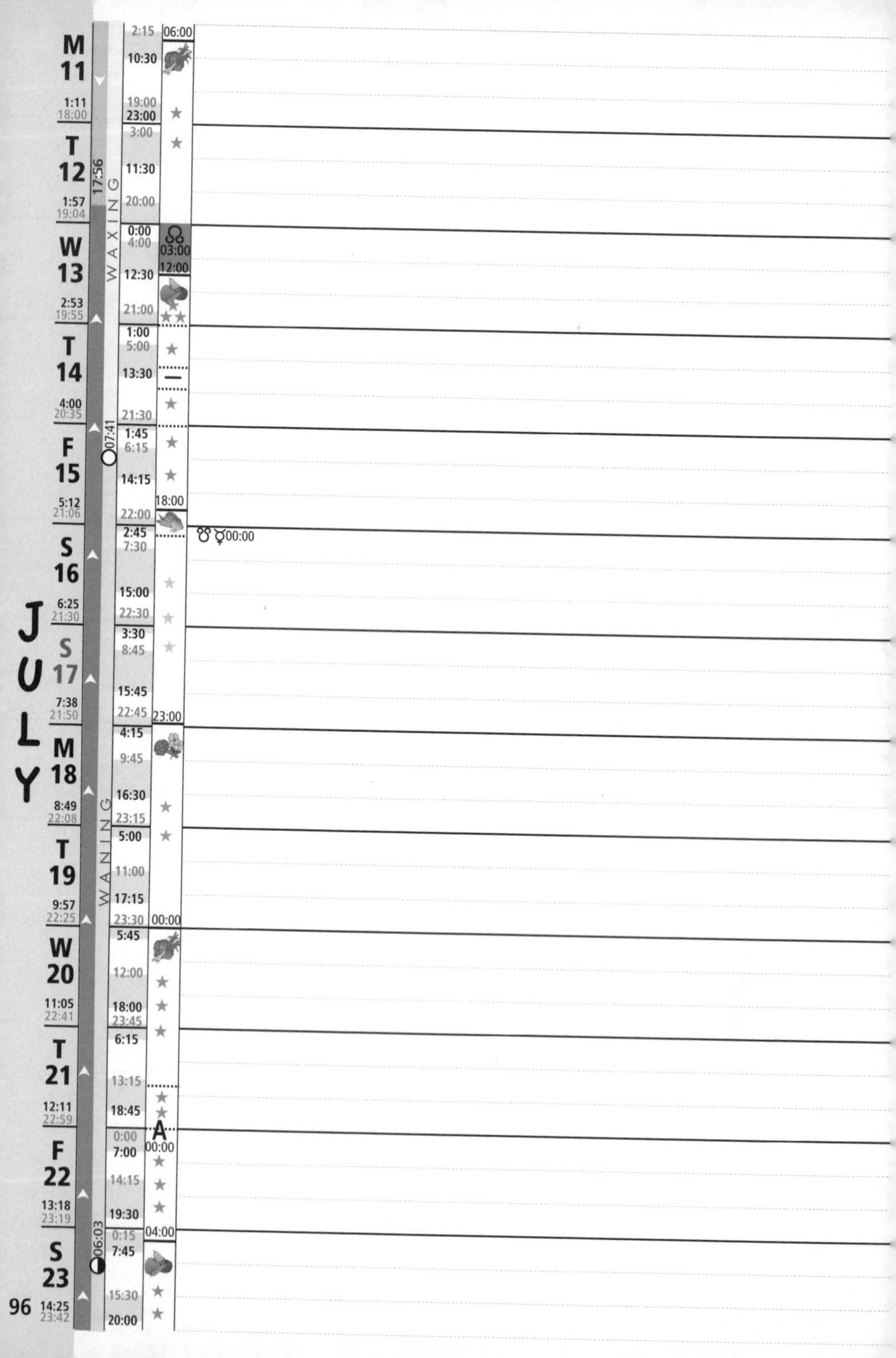

JULY

Day	Times	Periods	Notes
M 11	1:11 / 18:00	2:15, 10:30, 19:00, 23:00	06:00
T 12	1:57 / 19:04	3:00, 11:30, 20:00	17:56
W 13	2:53 / 19:55	0:00, 4:00, 12:30, 21:00	☊ 03:00, 12:00
T 14	4:00 / 20:35	1:00, 5:00, 13:30, 21:30	
F 15	5:12 / 21:06	1:45, 6:15, 14:15, 22:00	○ 07:41; 18:00
S 16	6:25 / 21:30	2:45, 7:30, 15:00, 22:30	☋ ☿ 00:00
S 17	7:38 / 21:50	3:30, 8:45, 15:45, 22:45	23:00
M 18	8:49 / 22:08	4:15, 9:45, 16:30, 23:15	
T 19	9:57 / 22:25	5:00, 11:00, 17:15, 23:30	00:00
W 20	11:05 / 22:41	5:45, 12:00, 18:00, 23:45	
T 21	12:11 / 22:59	6:15, 13:15, 18:45	A
F 22	13:18 / 23:19	0:00, 7:00, 14:15, 19:30	00:00; 04:00
S 23	14:25 / 23:42	0:15, 7:45, 15:30, 20:00	◐ 06:03

WAXING

WANING

JULY
AUGUST
S 24
M 25
T 26
W 27
T 28
F 29
S 30
S 31
M 1
T 2
W 3
T 4
F 5
WANING
WAXING
19:41
04:03
04:00
13:00
01:00
02:00
14:00
P
22:00
00:00

AUGUST
S 6
14:35
23:13
12:09
7:30
15:30
20:00
05:00
S 7
15:50
23:55
0:15
8:30
16:45
21:00
12:00
M 8
16:56
1:00
9:30
18:00
21:45
00:21
T 9
0:47
17:51
1:45
10:15
18:45
22:45
08:00
18:00
WAXING
W 10
1:49
18:34
2:45
11:15
19:30
23:45
T 11
2:58
19:07
4:00
12:15
20:00
01:00
F 12
4:10
19:34
0:30
5:15
13:00
20:30
S 13
5:22
19:55
19:59
1:30
6:15
13:45
21:00
S 14
6:33
20:14
2:15
7:30
14:30
21:15
07:00
M 15
7:42
20:31
2:45
8:45
15:15
21:30
T 16
8:50
20:48
3:30
9:45
16:00
21:45
08:00
WANING
W 17
9:57
21:05
4:15
11:00
16:30
22:00
T 18
11:04
21:24
5:00
12:00
17:15
22:30
17:00
A

Day	Date	Moon times	Times	Notes
F	19	12:10 / 21:46	5:45, 13:15, 18:00, 22:45	12:00
S	20	13:16 / 22:13	6:30, 14:15, 18:45, 23:15	
S	21	14:20 / 22:46	7:15, 15:15, 19:30, 23:45	12:00; 22:56 ◑
M	22	15:21 / 23:28	8:00, 16:15, 20:30	
T	23	16:17	0:30, 8:45, 17:15, 21:15	13:21; ♉ 18:00
W	24	0:20 / 17:05	1:15, 9:45, 18:00, 22:15	10:00
T	25	1:23 / 17:44	2:30, 10:45, 18:45, 23:00	—
F	26	2:35 / 18:17	3:30, 11:30, 19:15	12:00
S	27	3:54 / 18:45	0:00, 5:00, 12:30, 19:45	00:00
S	28	5:16 / 19:09	1:00, 6:15, 13:15, 20:15	04:05 ●
M	29	6:40 / 19:31	1:45, 7:45, 14:15, 20:30	
T	30	8:05 / 19:53	2:45, 9:00, 15:00, 21:00	09:00; P 19:00
W	31	9:31 / 20:16	3:30, 10:30, 16:00, 21:15	

WANING
WANING
WAXING

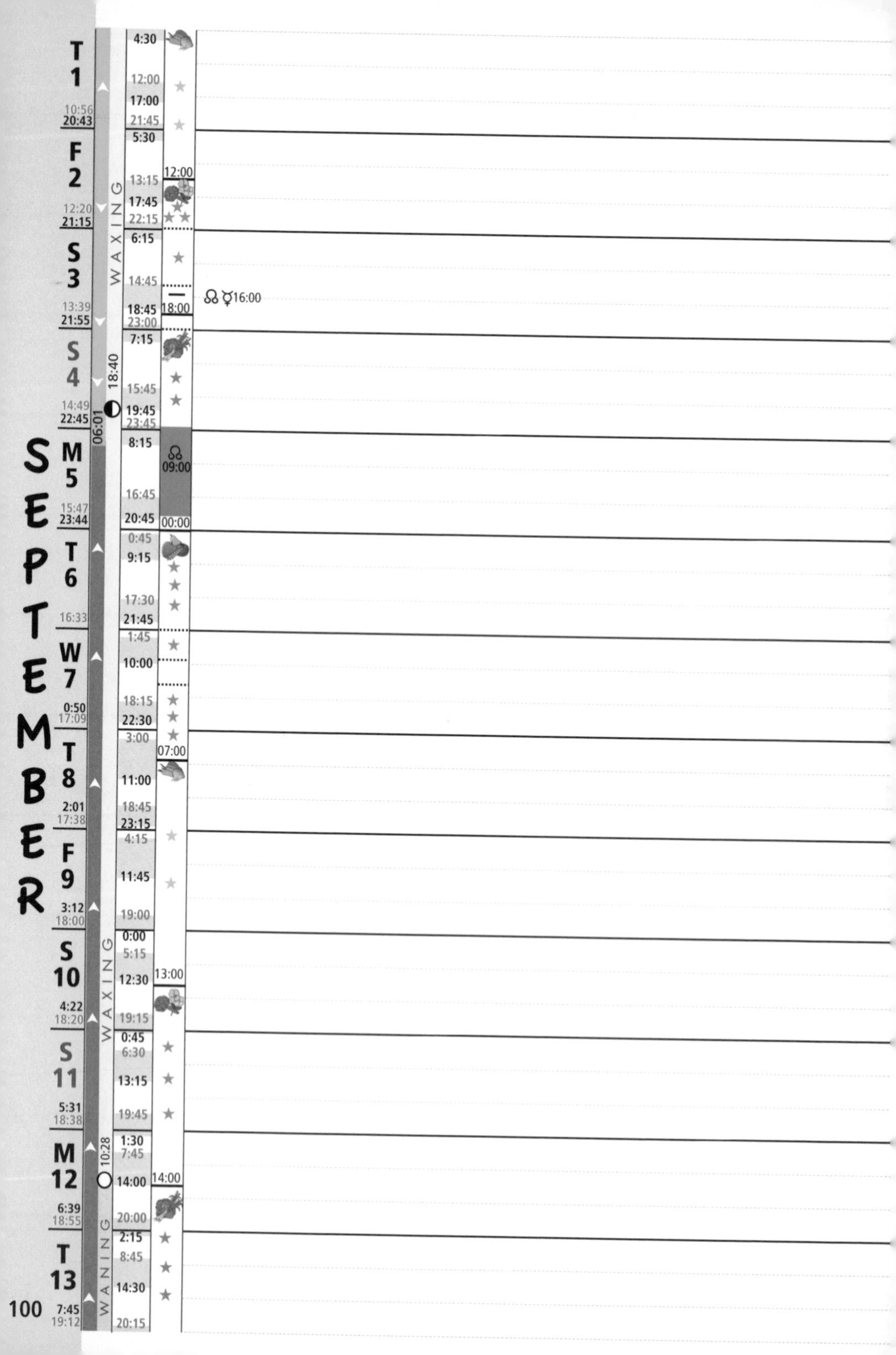

SEPTEMBER

Day	Moonrise / Moonset	Times	Notes
T 1	10:56 / 20:43	4:30, 12:00, 17:00, 21:45	WAXING
F 2	12:20 / 21:15	5:30, 13:15, 17:45, 22:15	12:00
S 3	13:39 / 21:55	6:15, 14:45, 18:45, 23:00	18:00; ☊ ☿16:00
S 4	14:49 / 22:45	7:15, 15:45, 19:45, 23:45	◐ 18:40; 06:01
M 5	15:47 / 23:44	8:15, 16:45, 20:45	☊ 09:00; 00:00
T 6	16:33	0:45, 9:15, 17:30, 21:45	
W 7	0:50 / 17:09	1:45, 10:00, 18:15, 22:30	
T 8	2:01 / 17:38	3:00, 11:00, 18:45, 23:15	07:00
F 9	3:12 / 18:00	4:15, 11:45, 19:00	
S 10	4:22 / 18:20	0:00, 5:15, 12:30, 19:15	13:00; WAXING
S 11	5:31 / 18:38	0:45, 6:30, 13:15, 19:45	
M 12	6:39 / 18:55	1:30, 7:45, 14:00, 20:00	○ 10:28; 14:00
T 13	7:45 / 19:12	2:15, 8:45, 14:30, 20:15	WANING

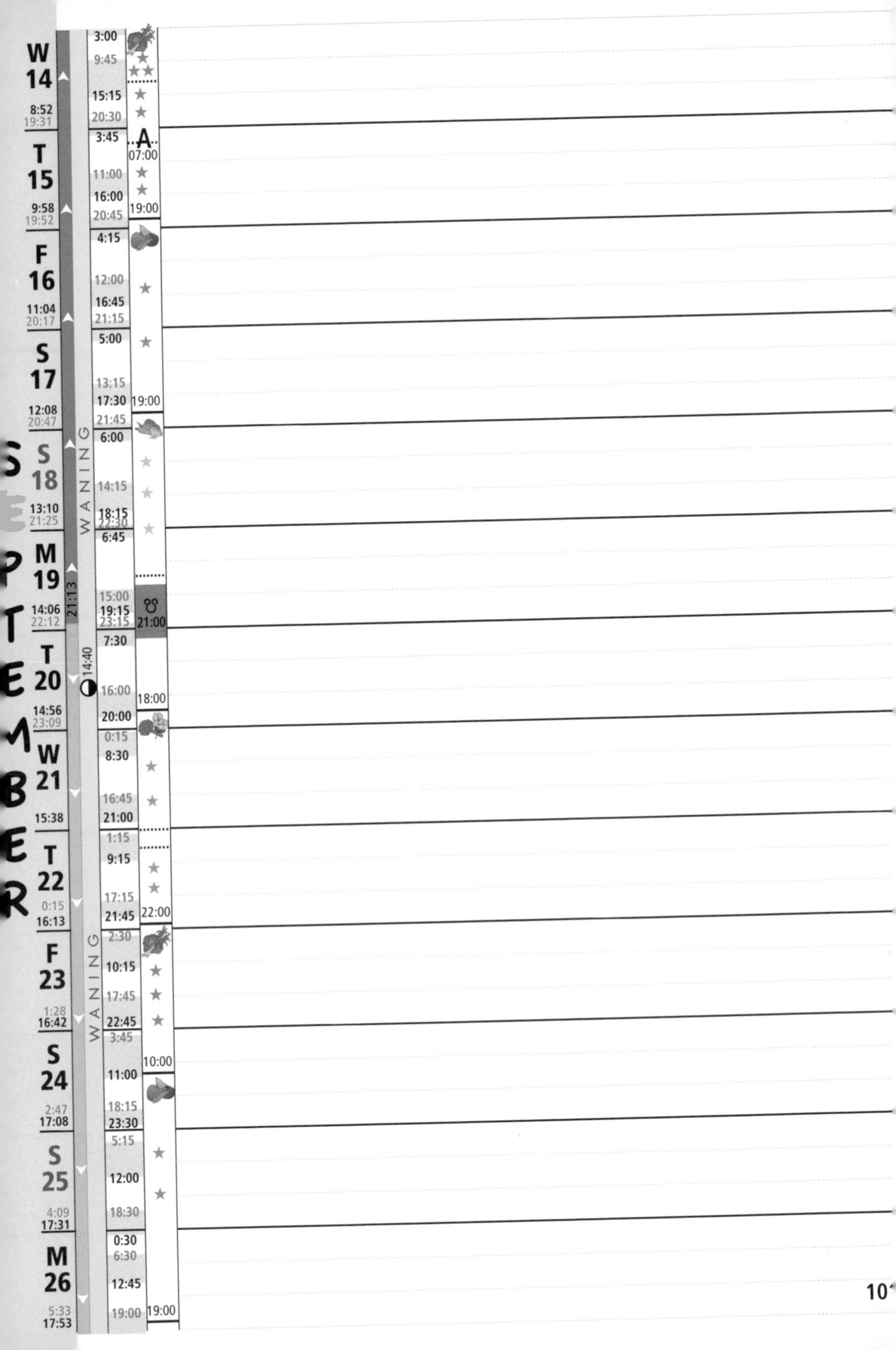

SEPTEMBER
W 14
8:52
19:31
3:00
9:45
15:15
20:30
T 15
9:58
19:52
3:45
A
07:00
11:00
16:00
20:45
19:00
F 16
11:04
20:17
4:15
12:00
16:45
21:15
S 17
12:08
20:47
5:00
13:15
17:30
21:45
19:00
S 18
13:10
21:25
6:00
14:15
18:15
22:30
WANING
M 19
14:06
22:12
6:45
15:00
19:15
23:15
21:13
21:00
T 20
14:56
23:09
7:30
14:40
16:00
20:00
18:00
W 21
15:38
0:15
8:30
16:45
21:00
T 22
0:15
16:13
1:15
9:15
17:15
21:45
22:00
F 23
1:28
16:42
2:30
10:15
17:45
22:45
WANING
S 24
2:47
17:08
3:45
11:00
18:15
23:30
10:00
S 25
4:09
17:31
5:15
12:00
18:30
M 26
5:33
17:53
0:30
6:30
12:45
19:00
19:00

SEPTEMBER
T 27
6:59
18:17
12:10
1:15
8:00
13:45
19:15
P
02:00
W 28
8:27
18:43
2:15
9:30
14:45
19:45
T 29
9:54
19:14
3:15
11:00
15:45
20:15
20:00
F 30
11:18
19:52
WAXING
4:15
12:15
16:45
20:45
01:00
OCTOBER
S 1
12:34
20:39
5:15
13:30
17:45
21:45
S 2
13:39
21:37
12:33
6:15
14:45
18:30
22:30
☊
10:00
M 3
14:30
22:42
7:00
15:30
19:30
23:45
06:00
T 4
15:10
23:52
04:16
8:00
16:15
20:30
W 5
15:41
0:45
9:00
16:45
21:15
13:00
T 6
1:03
16:05
WAXING
2:00
9:45
17:00
22:00
F 7
2:14
16:26
3:15
10:30
17:30
22:45
19:00
S 8
3:22
16:44
4:15
11:15
17:45
23:30
S 9
4:30
17:02
5:30
11:45
18:00
20:00

OCTOBER
M 10
0:15
6:30
12:30
18:15
5:36
17:19
WAXING
T 11
1:00
7:45
13:15
18:30
6:42
17:37
03:07
♉☿ 23:00
W 12
1:30
8:45
14:00
A
13:00
19:00
01:00
7:48
17:57
T 13
2:15
10:00
14:45
19:15
8:54
18:21
F 14
3:00
11:00
15:30
19:45
01:00
9:59
18:50
WANING
S 15
3:45
12:00
16:15
20:30
11:01
19:25
S 16
4:45
13:00
17:00
21:15
♉
22:00
11:59
20:09
03:12
M 17
5:30
13:45
18:00
22:00
01:00
12:51
21:02
T 18
6:15
14:30
18:45
23:00
13:34
22:03
W 19
7:15
15:15
19:45
06:00
14:11
23:12
04:31
T 20
0:15
8:00
15:45
20:30
14:41
F 21
1:30
9:00
16:15
21:15
19:00
0:25
15:08
S 22
2:45
9:45
16:30
22:15
1:43
15:31
WANING

OCTOBER
S 23
3:03
15:53
4:00
10:30
17:00
23:00
M 24
4:26
16:15
5:30
06:00
11:30
17:15
T 25
5:51
16:40
0:00
6:45
12:15
17:45
WANING
W 26
7:19
17:08
0:45
8:15
13:15
18:15
20:57
P
13:00
☋ ♀ 01:00
T 27
8:46
17:43
1:45
07:00
9:45
14:15
18:45
F 28
10:08
18:28
2:45
11:15
12:00
15:15
19:30
S 29
11:21
19:23
21:10
3:45
12:15
16:15
20:30
S 30
11:20
19:28
4:00
12:15
14:00
16:30
20:30
M 31
12:06
20:39
4:45
13:00
17:15
21:45
NOVEMBER
T 1
12:41
21:51
5:45
–
13:45
18:15
19:00
22:45
WAXING
W 2
13:08
23:03
6:30
16:39
14:15
19:00
T 3
13:31
0:00
7:30
14:30
19:45
00:00
F 4
0:13
13:50
1:15
8:15
14:45
20:30
WAXING
S 5
1:21
14:08
2:15
8:45
15:15
21:15

NOVEMBER

WAXING

Day	Moonrise / Moonset	Times	
S 6	2:27 / 14:25	3:30, 9:30, 15:30, 22:00	01:00 ★
M 7	3:33 / 14:43	4:30, 10:15, 15:45, 22:30	★ ★
T 8	4:39 / 15:03	5:45, 11:00, 16:00, 23:15	A 13:00 ★
W 9	5:45 / 15:25	6:45, 11:45, 16:30	06:00 ★
T 10	6:50 / 15:53	0:00, 7:45, 12:30, 17:00	○ 20:17 ★ ★ ★
F 11	7:54 / 16:26	0:45, 9:00, 13:15, 17:30	06:00 ★
S 12	8:53 / 17:08	1:30, 10:00, 14:00, 18:15	★ ★
S 13	9:47 / 17:58	2:30, 10:45, 14:45, 19:00	07:24; 01:00 ☊ ★ ★ ★
M 14	10:33 / 18:57	3:15, 11:30, 15:45, 20:00	06:00
T 15	11:12 / 20:03	4:15, 12:15, 16:30, 21:00	★ ★
W 16	11:44 / 21:13	5:00, 12:45, 17:30, 22:15	11:00 ★
T 17	12:10 / 22:27	5:45, 13:15, 18:15, 23:30	★ ★ ★; 01:00
F 18	12:34 / 23:44	6:45, 13:30, 19:00	◑ 15:10 ★ ★ ★
S 19	12:55	0:45, 7:30, 14:00, 19:45	★ ★

WANING

NOVEMBER
S 20
1:02
13:17
2:00
8:15
14:15
20:45
14:00
M 21
2:23
13:39
3:30
9:15
14:45
21:30
T 22
3:46
14:04
4:45
10:00
15:00
22:30
W 23
5:12
14:35
6:15
11:00
15:30
23:30
17:00
23:00
WANING
T 24
6:36
15:14
7:30
12:00
16:15
P
22:00
06:11
F 25
7:55
16:04
0:30
9:00
13:00
17:00
06:39
S 26
9:02
17:05
1:30
10:00
14:00
18:00
01:00
☊
00:00
S 27
9:56
18:16
2:30
11:00
15:00
19:15
M 28
10:37
19:30
3:30
11:30
16:00
20:30
04:00
T 29
11:09
20:45
4:30
12:15
16:45
21:45
WAXING
W 30
11:34
21:58
5:15
12:30
17:45
23:00
☊ ☿14:00
DECEMBER
T 1
11:54
23:08
6:00
13:00
18:30
08:00
F 2
12:13
0:15
6:45
13:15
19:15
09:53
S 3
0:16
12:31
1:15
7:30
13:30
19:45
08:00

DECEMBER
S 4
1:22
12:48
2:15
8:15
13:45
20:30
M 5
2:28
13:07
3:30
9:00
14:00
21:15
01:00
T 6
3:34
13:29
4:30
9:30
14:30
22:00
A
13:00
W 7
4:39
13:55
5:45
10:15
15:00
22:45
T 8
5:44
14:26
6:45
11:00
15:30
23:30
13:00
F 9
6:45
15:05
7:45
12:00
16:00
WAXING
S 10
7:42
15:52
13:37
14:37
0:15
8:45
12:45
16:45
07:00
S 11
8:31
16:49
1:15
9:30
13:45
17:45
12:00
M 12
9:13
17:54
2:00
10:15
14:30
19:00
T 13
9:47
19:04
3:00
10:45
15:15
20:00
17:00
W 14
10:15
20:18
3:45
11:15
16:15
21:15
07:00
WANING
T 15
10:39
21:33
4:30
11:45
17:00
22:30
F 16
11:01
22:49
5:30
12:00
17:45
23:45
S 17
11:22
6:15
12:15
18:30
20:00

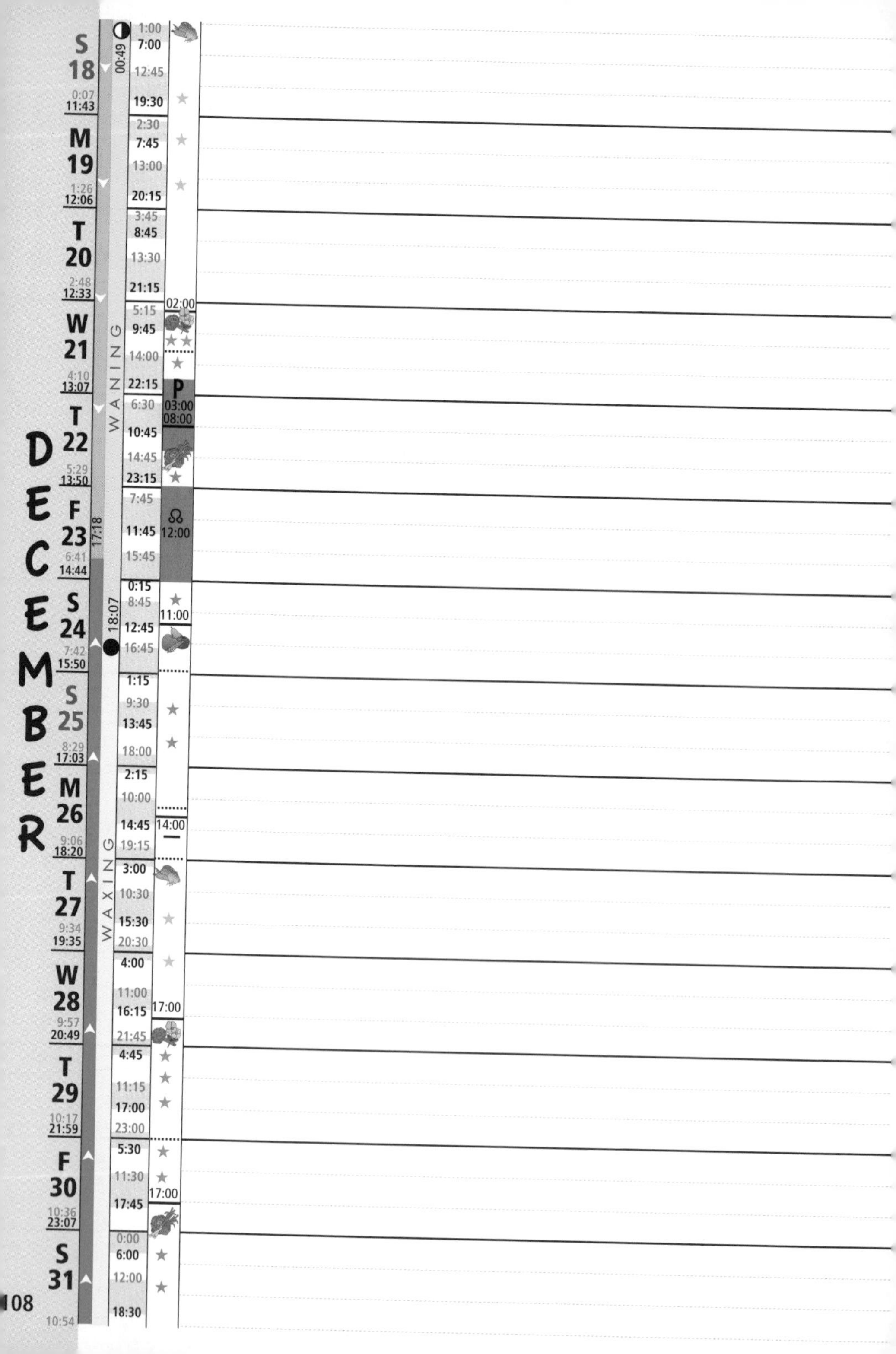

DECEMBER

Day	Date	Times	Time column	Markers
S	18	0:07 / 11:43	1:00, 7:00, 12:45, 19:30	◑ 00:49
M	19	1:26 / 12:06	2:30, 7:45, 13:00, 20:15	
T	20	2:48 / 12:33	3:45, 8:45, 13:30, 21:15	
W	21	4:10 / 13:07	5:15, 9:45, 14:00, 22:15	02:00
T	22	5:29 / 13:50	6:30, 10:45, 14:45, 23:15	P 03:00 08:00
F	23	6:41 / 14:44	7:45, 11:45, 15:45	☊ 12:00; 17:18
S	24	7:42 / 15:50	0:15, 8:45, 12:45, 16:45	11:00; ● 18:07
S	25	8:29 / 17:03	1:15, 9:30, 13:45, 18:00	
M	26	9:06 / 18:20	2:15, 10:00, 14:45, 19:15	14:00
T	27	9:34 / 19:35	3:00, 10:30, 15:30, 20:30	
W	28	9:57 / 20:49	4:00, 11:00, 16:15, 21:45	17:00
T	29	10:17 / 21:59	4:45, 11:15, 17:00, 23:00	
F	30	10:36 / 23:07	5:30, 11:30, 17:45	17:00
S	31	10:54	0:00, 6:00, 12:00, 18:30	

WANING

WAXING

Plan of your garden

Companion planting

Plants release secretions into the soil via their roots so one plant can affect the growth of another nearby. The Chinese and the South American Indians have known about this and made good use of the knowledge for centuries. Organizing your garden so that you position those plants that have a beneficial effect on each other side by side is another way to help improve your gardening success. Try your own companion planting experiments; results will vary according to the type of soil.

Plant	Grows well with	Does not grow well with
Asparagus	Cucumber, leek, parsley, pea, tomato	Beetroot, garlic, onion
Aubergine	French bean	Potato
Beetroot	Cabbage, celery, lettuce, onion	Asparagus, carrot, leek, runner bean, tomato
Cabbage (except kohlrabi)	Beetroot, celery, corn salad, French bean, lettuce, onion, pea, potato, tomato	Chicory, fennel, garlic, leek, parsnip, radish
Carrot	Chervil, chive, dwarf French bean, lettuce, onion, leek, parsley, parsnip, pea, radish, tomato	Beetroot, Swiss chard
Celery	Beetroot, cabbage, cucurbitaceous plants, French bean, leek, pea, Swiss chard, tomato	Lettuce, maize, parsley
Chicory		Cabbage
Corn salad	Cabbage, leek, onion, strawberry	
Cucumber	Asparagus, basil, cabbage, celery, chive, French bean, lettuce, maize, pea	Potato, radish, tomato
Fennel	Celery, leek tomato	Cabbage, French bean, parsnip,
French bean	Aubergine, carrot, celery, cabbage, lettuce, maize, marigold, potato, spinach, radish, strawberry, turnip	Beetroot, garlic, fennel, onion, shallot, Swiss chard
Garlic	Potato, strawberry	Cabbage, French bean, marigold, pea

Plant	Grows well with	Does not grow well with
Kohlrabi	Beetroot, celery, leek, lettuce	Chicory, fennel, radish, strawberry
Leek	Asparagus, carrot, celery, fennel, lettuce, onion, strawberry, tomato	Beetroot, cabbage, parsley, pea, Swiss chard
Lettuce	Beetroot, cabbage, carrot, chervil, cucumber, broad bean, leek, onion, pea, strawberry, radish, turnip	Parsley
Maize	Cucurbitaceous plants, French beans, pea, potato, tomato	Beetroot, celery
Marrow and courgette	Basil, maize, nasturtium, potato	Potato, radish
Onion	Beetroot, carrot, cucumber, leek, lettuce, parsley, parsnip, strawberry, tomato	Cabbage, French bean, pea, potato
Parsnip	Carrot, onion	Fennel
Pea	Asparagus, cabbage, carrot, celery, cucumber, lettuce, maize, potato, radish, turnip	Garlic, leek, onion, parsley, shallot
Potato	Broad bean, cabbage, celery, French bean, garlic, marigold, nasturtium, pea, tomato	Aubergine, cucumber, maize, onion
Radish	Carrot, chervil, French bean, garlic, lettuce, pea, spinach, tomato	Cabbage, marrow
Spinach	Cabbage, French bean, radish, strawberry, turnip	Beetroot, Swiss chard
Strawberry	Chive, corn salad, French bean, garlic, leek, lettuce, marigold, onion, spinach, thyme, turnip	Cabbage
Swiss chard	Celery, lettuce, onion	Asparagus, basil, leek, tomato
Tomato	Asparagus, basil, carrot, garlic, cabbage, celery, leek, French marigold, nasturtium, marigold, onion, parsley, parsnip, potato, radish	Beetroot, fennel, kohlrabi, pea, Swiss chard
Turnip	French bean, lettuce, pea, spinach	

Index

Animals: 19, 36, 37
Asparagus: 43,60

Beer: 40
Bees: 38
Blackcurrant bushes: 26
Boats (painting): 42
Bottling: 3, 40
Brambles: 34
Bread, sourdough: 43
Bulbs: 24
Bushes: 25, 26

Callouses: 49
Cattle: 37
Cereals: 35
Chicks: 37
Chimneys: 42
Christmas tree: 42
Chrysanthemums: 71
Cider: 40
Cladding: 41
Cleaning: 36, 43
Clearing land: 34
Clothes: 43
Comfrey: 29, 31
Compost: 3, 6, 16, 28, 30, 31, 51, 73
Conifers: 25, 26
Corns: 49
Cows: 37
Cowshed: 36, 37
Cuttings: 26, 71, 75
Convolvulus: 33
Covering (animals): 36
Cryptogenic diseases: 28, 30

Decanting: 40
Decoctions: 29
Depilation: 47
Dew: 27, 40, 48
Digestion: 50
Disinfecting, cowshed: 36
Disbudding: 39, 70
Drainage, soil: 42

Earthing up: 22
Endives: 20
Ewes: 36

Fasting: 49
Fertilizers: 24, 29, 30
Flowers: 10, 21, 24, 25, 45
Food: 43, 50
Framework, wood for: 41
Fruit: 5, 10, 20, 25, 26, 27
Fungicides: 30

Grafting: 5, 16, 24, 26, 62
Grapes, cultivating: 39
Green manure: 22, 32, 62

Hair: 46, 47
Harvesting: 7, 16, 17, 27, 39, 45, 72
Hay: 34
Hedges: 26
Hemisphere: 3, 4, 5, 58
Hoeing: 20
Honey: 38
Horsetail: 22,23, 29, 30, 66

Infusions: 45
Insecticides: 30

Jams: 3, 43

Lawn: 5, 6, 43
Leaves: 15, 19, 26
Liquid fertilizer: 29, 30

Manure: 31, 32, 34
Mares: 36
Medicinal plants: 45
Moon ascending-descending: 4, 5, 7, 25, 29, 54, 55
Moon waxing-waning: 2, 3, 4, 5, 25, 29
Moss, mould: 42
Mulching: 20

Nails: 49

Paths: 42
Planting: 20, 25, 29, 54
Ploughing: 33
Poplars: 25
Potatoes: 1, 22, 63
Preserving: 3, 43, 73, 76
Pricking out: 6, 17, 20, 21, 33,
Pruning: 6, 24, 25, 39, 57, 60

Rabbits: 36
Repotting: 24
Roots: 6, 10, 20, 21, 25, 27
Rosebushes: 24

Salad: 21
Sap (silver birch): 5, 43
Sauerkraut: 43
Shallots: 27, 60
Shoeing: 36
Shrubs: 25, 26
Skin: 48
Soil: 7, 15, 16, 29, 31, 33, 54
Sowing: 7, 16, 17, 20, 55, 64
Springs: 42
Squashes: 27
Stinging nettles: 29, 30, 31
Strawberry plants: 20

Thistles: 33
Thorns: 34
Tides (influences of): 5, 16, 17, 27
Tomatillos: 23
Treatments: 48, 49
Trees: 19, 25, 26
Trenches: 42
Trimming feet, livestock: 36

Warts: 48
Watering: 20, 25, 69
Weeds: 33
Wine: 3, 39
Wood: 6, 17, 40, 41, 42, 79
Worms: 31, 49